ELECTRICAL AND ELECTRONIC TECHNOLOGIES:

A Chronology of Events and Inventors from 1940 to 1980

by
HENRY B. O. DAVIS

The Scarecrow Press, Inc.
Metuchen, N.J., & London
1985

Library of Congress Cataloging in Publication Data

Davis, Henry B. O., 1911–
  Electrical and electronic technologies.

  Bibliography: p.
  Includes index.
  1. Electric engineering--Chronology. 2. Electronics
--Chronology. 3. Electricity--Chronology. I. Title.
TK15.D375 1985   621.3'09'04   84-13957
ISBN 0-8108-1726-8

To my six Davis grandchildren in chronological order ...
    naturally:

ELIZABETH
SUSAN
JOHNATHAN
ELAINE
JORDAN
PAMELA

and those yet to come....

CONTENTS

PREFACE

As with the previous two volumes of this series, it is the purpose of this book to provide students, teachers and authors with a ready source of information on specific events in the development of the electrical and electronic technologies in a year-by-year progression.

The first volume covers the period from the earliest times to 1900. This was followed by a second volume, which begins in 1900 and covers the years to 1940.

This third volume covers the period from 1940 to 1980. This was a time that saw the transistor take over the duties of the vacuum tube, which was soon replaced by integrated circuits. This later development brought about seemingly impossible size reductions in nearly all electronic devices.

It was the period of the Great War, which spurred the development of new systems and devices and supported research in many fields. Many new electronic instruments were developed to support research in programs not related to electronics or electricity.

It was the period when man entered the space age, walked on the moon, photographed the planets and their moons, and made soft landings on the surfaces of Mars and Venus.

It was the period during which the computer industry grew from a few companies making very large computers to many companies making smaller and smaller computers as integrated circuit techniques were applied.

It was a period which brought spin-offs from laser technology with applications in medicine, computer memories, measurements and other breakthroughs for the betterment of humanity.

The story is summarized at the beginning of each decade. The format follows that of the previous volumes. In general, paragraphs dealing with the same subject area such as "power," "television," "amateur," and so forth will be grouped in adjacent paragraphs in any given year.

Chapter 1

THE FIFTH DECADE: 1940-1949

The fifth decade of the century, the decade of the Second World War, was a period of progress. Hindered in some areas of development because of war priorities, the development in areas relevant to the war and national defense was indeed impressive.

One group not making tremendous gains early in the decade was the radio amateur. If you were a ham and turned on your receiver on December 7, 1941, you were no doubt surprised at the lack of signals on the band. Tuning from one end of the band to the other on the 40-meter band, there was nothing but a few weak tweets. After calling CQ, the general call to all stations, you immediately got a call saying "the bands are closed--get off the air."

Japan had attacked Pearl Harbor. Only a few stations were on the air and they were there to help clear the bands and for civil defense purposes. Although the bands were all closed, all was not lost. It is reported that the Yagi beam antenna was developed about 1941 in an effort to make a death ray. It did not serve for that purpose, but it made a very good beam antenna and was widely used in later years.

Although the radio waves could not be used by the amateur stations, other means of communication were experimented with, including wired wireless, light beams, inductive fields and ground conduction using two ground rods. Undoubtedly the most significant event for the amateur was the introduction of single sideband suppressed carrier (SSSC). The improvement in single sideband was decidedly apparent in the reduced interference from adjacent stations in the band, and the effective power increase by suppressing one sideband and the carrier. SSB was rapidly accepted by the phone men (amateurs preferring voice to code communication) and, despite its increased complexity, sounded the death knell for conventional amplitude modulation in the ham bands. A number of distance records were made in the VHF and UHF bands and the first amateur TV stations came on the air during this decade.

On January 9, 1942, all amateur stations including those that had been used for civil defense purposes were closed down for the duration of the war. The Signal Corps needed radio equipment and stations and parts were purchased from the hams. Other hams in Signal Corps repair shops were busy building or revising transmitters and equipment to meet military needs, or building transmitters and other equipment for the military with the parts purchased from the amateurs. By the end of November 1946, all amateur bands were again open for operation.

The decade was a time of experimenting with frequency modulation (FM). The Federal Communications Commission (FCC) authorized FM broadcasting to start in 1941 in the 42 to 50 MHz (megahertz) band. At that time the call letters of the station indicated the frequency of transmission with the last letter indicating the city of origin. As the number of stations increased, it became evident that such a system was not satisfactory. An all-letter system such as is used with the AM stations came next. By the middle of the decade, so many companies were getting two sets of call letters that the FCC again changed its call system. The new method for companies having an AM station was to use the same call as the AM station followed by FM.

The demand for frequencies and the problem of sky-wave interference again made a change necessary. In 1945, the FCC moved FM broadcasting to the 88 to 108 MHz band. This band provided more channel space and minimized the sky-wave problem. FM was well received by the public and by 1945 there were over 350 applications for licenses on file at the FCC.

By 1940 there were 777 amplitude modulated (AM) stations in operation in this country with an estimated 500,000,000 receiving sets in use. A number of broadcasting chains for AM and at least one for FM broadcasting were in operation. The National Broadcasting Company (NBC) owned two chains, the Red and Blue networks. With the FCC ruling that no more than one network could be controlled by one organization, NBC was forced to sell one of its networks. The Blue network was sold and became the American Broadcasting Company (ABC).

Radio came into use as a means of assisting in marine navigation. Among these systems, Loran, Shoran, Gee and Decca stations were in operation, being built, or in the testing phase. Size reduction of radio equipment, both in space and weight, was a necessity because of the military. Walkie-talkie transceivers, easily carried in one hand, were developed. Even after the war, the trend toward smaller and lighter equipment continued. Complete receiving sets, using small vacuum tubes, were made small enough to fit in a shirt pocket, including the antenna and speaker.

During the decade, the market for AM receivers had become about saturated. To jack up sales, the clock radio came onto the market. The clock was usually combined with a five-tube superheterodyne AC-DC radio. The age of the large console radio was about gone and the console television or AM-phonograph-TV was taking its place. High fidelity and stereophonic sound was being pushed by record manufacturers. Stereo broadcasting started both in this country and abroad.

Trains, themselves, appeared to be on the way out. Some were trying to improve service by providing recorded music in some of its cars and rooms. VHF was used for communicating from the cab to the rear on some freight trains.

The military applications of electronics were pushed with remote control experiments and application in torpedoes, ships, and aircraft. Development of the proximity fuse was pushed to a successful conclusion and put in use against the buzz bomb and aircraft in Europe and the Pacific.

A number of instrumentation systems were developed primarily for the military; the sniperscope, radar altimeter, and antiaircraft gun control, to name a few. The war emergency also accelerated the development of radar with high-power magnetrons permitting higher resolution with smaller

antenna systems. Radars containing the plan position indicator were installed on many ships providing nearly continuous observation in all directions, making a sneak aircraft attack almost impossible.

The field of radio astronomy was in its infancy at the
start of the decade. The interest in the field had declined
in this country but was revived after a paper was published
by Grote Reber in 1940. By the middle of the year, he had
definitely verified that the sun was indeed a noise source as
had been reported several years earlier by J.S. Hey.

The idea of reaching the moon by radar had been suggested earlier and this was accomplished in 1946, another
shot in the arm for radio astronomy. Radio telescopes were
set up at a number of institutions in this country and
abroad, and the search began for indications of electromagnetic radiation from the planets. At the end of the decade
it was discovered that some stars were very intense radia-
tors. Facsimile had been developed to where commercial
radiophoto service was available between the US and a number of foreign countries.

A hold was put on television licensing by the FCC in order for acceptable standards to be agreed on by the various
manufacturers. All electronic and electromechanical systems
were being pushed by their developers. The Radio Manufacturers Association set up the National Television System
Committee (NTSC) to formulate standards acceptable to the
industry. The standards were submitted to the FCC and
were officially adopted. Later the FCC set up a subcommittee on color television. Color TV was demonstrated but TV
development was stopped because of war priority. When TV
activities started after the war, RCA brought out its first
commercial TV, a 10" black-and-white set. By 1948, the
FCC ceased issuing color TV licenses until definite color
standards could be agreed on. Various demonstrations were
held. RCA demonstrated color pictures on 8' x 10' screens,
but these large pictures were generally not bright enough to
be clearly seen in the average ambient light of the home.
Some small, lightweight television cameras were developed to
permit monitoring the meters in drone aircraft and thus serve
as a telemetering system.

With the possibility of war on the horizon as the decade

opened, the trend toward smaller and lighter components
soon became a necessity for military equipment. One of the
most outstanding developments was the tube required for the
variable time fuse. This tube had to be both very small
and very rugged, capable, in fact, of withstanding shocks
on the order of 20,000 times the force of gravity when being
shot from a gun. These tubes were developed in time to
prove their effectiveness in combat. To go along with the
smaller tubes, other components were also reduced in size.
Metallized paper capacitors were brought out. Miniature
transformers, potentiometers, and other small devices were
developed.

In general, military equipment had to be small and light,
but reliability was the most vital factor. One of the factors
responsible for failure of electronic equipment was humidity,
causing failure by corrosion or leakage. To overcome this,
circuits were sprayed with a moisture-proofing varnish which
greatly improved reliability in the field operation. There
were, however, some components which could not be treated
this way, such as meters and relays. To solve this problem, hermetically sealed relays and meters soon appeared.

Along with the size and weight reduction came new fabrication techniques; stamped wiring, stamped from copper
sheets, was tried but soon replaced by printed circuits and
dip soldering, the soldering of many components to a printed
circuit board in one operation.

By the end of the decade, but not in time to aid the
military in WW II, came the development of the transistor and
the beginning of the era of solid state. The initial problem
with the transistor was caused by impurity in the germanium.
This problem limited the application of the transistor. Differ-
ences in the germanium caused by impurities made it almost
impossible to design a circuit which would work equally well
with another transistor of the same type. These problems
were largely solved in the 1950's.

The war did little to delay the development of the computer. During this decade computers developed, from the
relatively slow relay types that required up to a second for
multiplication, to the all-electronic Eniac that needed only
several milliseconds for the same operation. By the end of
the 1940's, the descendent of the Eniac was under construc-

tion. This was the Univac, an all-electronic computer with an internal memory and a clock rate of 2,250,000 PPS (pulses per second). Analog computers were also advanced by direct coupled amplifiers which made the all-electronic analog computers practical. Analog computers were used for gun directors during the war and for designing ships before this was feasible with digital computers.

Tremendous advances were made in atomic physics. A number of particle accelerators were developed such as the betatron, synchrotron, and linear accelerators which gave energies to atomic particles in excess of a billion electron volts. Work early in the decade showed that atomic fission was possible. With the demonstration that nuclear chain reaction was possible, because of the war emergency, work was started on the development of the atomic bomb. It was imperative that the United States get the bomb first. The success of this effort in 1945 brought a sudden end to the war and inflicted severe damage on several Japanese cities. Other progress was made in the knowledge of the structure of the atom and the magnetic properties of the atomic nuclei.

Electronics was also being used on boiler smokestacks such as those of the conventional power plant. Electronic monitors were used to control fuel combustion for more efficient combustion and a cleaner environment.

The voltage of transmission lines continued to climb. Voltages up to 380,000 were used. The multiconductor system to reduce corona loss was developed, and ACSR (aluminum conductor steel reinforced) was used on many lines because of the shortage of copper.

New developments continued to improve the telephone service with carrier current systems being more widely utilized. This system permitted a number of conversations to be carried simultaneously without cross talk. Coaxial cable use became more general with up to 600 telephone channels carried on one cable. Generally, one cable was used for one direction of transmission, while an adjoining cable carried the return communication. For the first time commercial telephone service was available to private vehicles. By 1950 this service was branching out to many of the larger cities.

<u>1940</u>

AMATEUR

1. <u>New Band</u>. As of April 13, part of the five-meter band was opened for amateur use but limited to frequency modulated (FM) voice transmission.

2. <u>Amateur Suspension</u>. At this time the war situation was getting extremely critical. On June 5, the Federal Communications Commission issued FCC order #72. Effective immediately, no amateur licensed by the FCC could communicate with any foreign country.

3. <u>Amateur Television</u>. So far as is known, the first two-way communication by television between amateurs occurred September 27 on 112 MHz between W2USA and W2DKJ/2. Voice contact was on 56 MHz. The demonstration of the equipment over an eight-mile path (approximately) was made at the Worlds Fair in New York and later at the Chicago Radio Show. Later in the year a new distance record for amateur TV was made by W2DKJ/2 and W2ADE over a 24-mile path.

BROADCASTING

4. <u>Frequency Modulation</u>. By fall 1939, the requests for frequency modulation building permits had so flooded the FCC that the issuing of the permits for new construction was temporarily discontinued to permit a study of the problem of getting more stations into the existing FM bands. On October 31, the FCC again began issuing construction permits. Commercial FM broadcasting was authorized to start January 1, 1941. Thirty-five channels were authorized for commercial use and five for educational programming.

5. <u>Zenith Radio Corporation</u>. On February 2, 1940, the Zenith Radio Corporation started its FM broadcasts over experimental station W9XEN. This was the first FM station in Chicago and one of the first in the country.

6. __AM-FM Sets__. With the increased interest in frequency modulation, manufacturers began to put on the market receivers capable of receiving either the conventional amplitude modulated stations or the new frequency modulated stations which were springing up all over the country.

7. __Tizard Commission__. About the end of August, a scientific mission from Britain headed by Sir Henry Tizard, Scientific Advisor to the Aircraft Ministry, came to Washington with the authority to trade British secret information for US secrets. Among the secrets disclosed to the United States was information on the resonant cavity magnetron. This device led to the development of high-power, high-frequency and high-definition radar suitable for "accurate bombing at night in spite of clouds and poor visual conditions."

COMMITTEES

8. __National Television System Committee__. The television industry had been in a confused state with at least three major developers competing for their standards to be adopted by the industry. The Radio Manufacturers Association had proposed standards but there was no complete agreement by the manufacturer with them. To produce standards independent of manufacturers' pressure the National Television System Committee was set up in March to review the recommendations and come up with a set of standards satisfactory to the industry.

9. __National Defense Research Committee__. Members of the National Advisory Committee on Aeronautics (NACA) had realized for some time that it was likely the United States would soon be drawn into the war. Even before the invasion of Poland by Germany it had become apparent to Dr. Vannevar Bush, Chairman of NACA, that the United States should have an organization for applying scientific principles to war. By the efforts of Bush, Karl T. Compton of MIT, the National Academy of Sciences, Bell Telephone Laboratories and others, the National Defense Research Committee (NDRC) was formed. The name was pro-

posed by John Victory, then Secretary of NACA.
The first regular meeting of NDRC was held July 2.

COMPONENTS

10. <u>Helipots</u>. Arnold Beckman invented the ten-turn po-
    tentiometer which he called the "Helipot." This device
    permitted hitherto unprecedented precision in setting
    continuously variable resistors.

11. <u>Non-Inductive Resistors</u>. A new line of non-inductive
    resistors, Series Z, was announced by the Clarostat
    Manufacturing Company. The resistors were rated
    from 10 to 100 w (watts) and were suitable for such
    purposes as dummy loads, rhombic antenna termina-
    tions and other high-frequency uses where power re-
    sistors were required.

12. <u>Tubes</u>. Some of the new tubes announced this year in-
    cluded: Eimac--152TL, 304TL; General Electric--
    8002R; Heintz and Kaufman--HK-257; Hytron--HY30Z,
    HY-75, HY 615; KenRad--1T5; Raytheon--1DB,
    6AB5/6N5, 6AL6G, 7B4, 7G7/1232, 7H7, 7J7, 7L7,
    35Z6G, 50C6G, 70L7GT, 117M7GT, RK 65; RCA--1R5,
    1S5, 1T4, 12K8, 12SR7, VR75-30, 811, 812, 815, 825,
    828, 829, 928, 1628, and the RCA-1847, a TV pickup
    tube; Sylvania--1LB4, 1LC5, 1LC6, 1LD5, 7B4, 7G7/
    1232, 7J7, 7L7; Western Electric--707A, 706Y.

COMPUTERS

13. <u>Calculator</u>. The first public display of remote data
    processing was made at Dartmouth College when the
    Stibitz Relay Computer and the Complex Number Cal-
    culator were demonstrated. Computing was done in
    New York with the input-output station at Dartmouth
    College in Hanover, New Hampshire.

14. <u>C.A. Lovell and D.B. Parkinson</u>. It was about this
    year that Lovell and Parkinson of the Telecommuni-
    cation Research Establishment developed the idea of
    using an analog computer for controlling antiaircraft
    fire.

CONFERENCE

15. Second Inter-American Radio Conference. The Second
Inter-American Radio Conference met January 18 to
January 25 in Santiago de Chile. All amateur bands
through 60 MHz were specified for amateurs only.
At this time no agreement was reached for subdivid-
ing the bands for radiotelephone use. The available
bands remained exclusively amateur throughout the
Americas.

ELECTRON MICROSCOPE

16. V.K. Zworykin. Dr. V.K. Zworykin of RCA Labs announced the completion of the electron microscope
with magnification of over 100,000 times (April 20).
It was on the market this year.

GUIDED MISSILE

17. Remote Control. By March, the possibility of war appeared more likely. The commander-in-chief of the
navy fleet directed that tests be conducted to determine the practicability of remote-controlled torpedoes.

ILLUMINATIONS

18. Dr. Harold A. Edgerton. The electronic flash lamp developed by Dr. Harold A. Edgerton at MIT was marketed this year.

19. Sealed-Beam Headlights. The sealed-beam headlight for
automobiles was developed to a form suitable for application. They were soon accepted for new automobile use.

INDUSTRY

20. Compagnie Française Thomson-Houston (CFTH). With
the merger of the Edison General Electric Company
and the Thomson-Houston Company in the 1890's, the

Thomson-Houston name was largely forgotten in this country. However, because of an agreement between the Thomson-Houston International Electric Corporation and a group of French industrialists, the Compagnie Française Thomson-Houston was organized with the aim of entering all fields of electricity. By 1940, after a period of mergers, it held a firm position in nearly all of the major electrical fields.

21. Hewlett-Packard Co. The Hewlett-Packard Co., first set up in the garage behind the Packards' house in Palo Alto, California (1938), had continued to grow to where it could no longer be contained in the Packards' garage. The company moved to a small building where it began to expand from the two-man organization. Business continued to increase, and in less than two years plans were under way for a new building.

INSTRUMENTATION

22. Signalyst. RCA announced a new signal generator of an unusually large frequency range, 120 kc to 120 mc.

23. Modulation Monitor. The Triplett Electrical Instrument Company announced a new modulation monitor which with a given carrier level reference would indicate the percentage of modulation directly. An overmodulation indicator would indicate if modulation exceeded a preset level within the range of 40 to 120 percent. (Model 1696K)

24. Direction Finder. Hallicrafters announced a direction finder utilizing a loop antenna and covering the range of 100 to 1,500 meters. This range covered the beacon, broadcast and marine bands.

25. Wavemeter. The General Radio Company introduced the type 758A UHF wavemeter operating from 55 to 400 MHz. This was an absorption type unit. Resonance was indicated by a lamp.

26. Instrumentation. A large number of new service instru-

ments for radio equipment were on the market this
year. Many models of multirange volt-ohm-current
meters were brought out. At least a dozen models
were advertised by three companies. Some were combined with tube testers. Other instruments to assist
service men were signal tracers, frequency meters,
tube testers and secondary frequency standards.

INTERFERENCE

27. Suppression. In Great Britain, a bill was introduced
which required every owner of an electrical machine
or automobile to install a suppressor for the electrical
noise it generated.

NAVIGATION AIDS

28. Robert J. Dippy. A system of radio navigation called
GEE was developed by Robert J. Dippy of the Telecommunication Research Establishment in England.
The system was somewhat similar to the Loran System
proposed in the United States.

29. Alfred Loomis. A scheme for a navigation aid was proposed in October by Alfred Loomis. This was a
scheme in which radio waves from shore stations could
be used for locating the position of a ship or aircraft.
The transmitters on land produced hyperbolic lines in
the form of a grid in which the position of the aircraft or ship could be determined from a set of charts
for a highly accurate fix. The scheme became known
as Loran (Long Range Aid to Navigation).

ORDNANCE

30. M.A. Tuve. The National Defense Research Committee
(NDRC) was requested to give top priority to the de-
velopment of proximity or influence fuses. In August,
Section T of NDRC was established under Dr. M.A.
Tuve of the Carnegie Institution for the development.
Later in the year, the Bureau of Standards joined in
the effort.

31. <u>Jan Forman</u>. Jan Forman, in Cambridge, England, developed a tube which would stand accelerations of about 15,000 G's suitable for the proximity fuse.

PHYSICS

32. <u>Donald W. Kerst</u>. Dr. Donald W. Kerst of the University of Illinois built the first practical betatron. A larger unit which he built later produced electron accelerations of up to 300 MeV. These high-energy electrons were used to develop extremely penetrating X rays.

RADAR

33. <u>Radar (the term)</u>. The term "radar," to indicate <u>RA</u>dio <u>D</u>etecting <u>A</u>nd <u>R</u>anging, had been credited to two naval officers, E.F. Furth and S.P. Tucker. It was considered possible that the term may have been independently conceived by both men. In November, the Chief of Naval Operations directed that the term be used in nonclassified material in reference to the then secret project.

34. <u>Marine Radar</u>. The first production-type radars made by RCA were the CXAM's installed on the battleship <u>California</u> and the cruisers <u>Chester</u>, <u>Chicago</u>, <u>North-ampton</u> and <u>Pensacola</u>. One was also installed on the aircraft carrier <u>Yorktown</u>.

35. <u>Radiation Laboratories</u>. Soon after the formation of the National Defense Research Committee (NDRC) it was decided that the quickest and best way to get into radar research would be through educational institutions where laboratories and scientific personnel were already available. The Massachusetts Institute of Technology was in an ideal location and had a group of men already working in the microwave field.

   In October MIT was chosen as the location for the microwave radar laboratory with Dr. Lee A. Du-Bridge as the director. Work started at the radiation laboratory on November 10 with a contract from NDRC. The organization grew from 30 or so employees at the start to a total of around 4,000 people. The first microwave pulse radar was built in about two months and successfully demonstrated on January 4, 1941.

RADIO

36. Pocket Radio. Pocket-size radios using miniature tubes came on the market this year. RCA introduced its "personal" receiver using subminiature tubes.

RADIO ASTRONOMY

37. Grote Reber. This year, Grote Reber wrote his first paper, "Cosmic Static," which received the notice of astronomers and revived interest in the field of radio astronomy, which had been more or less dormant for about three years.

38. John De Witt. On May 21, John De Witt indicated in his notebook the possibility of bouncing a UHF (ultra-high frequency) signal off the moon. The idea was to open up the study of the upper atmosphere and possibly world communication. De Witt was later in charge of Project Diana which succeeded in the effort.

RECORDING

39. Bias. A patent of Braunmuhl and Weber in Germany, covering high-frequency bias to the magnetophone magnetic tape, was the factor which allowed the magnetic recorder to reach the popularity it did because of the resulting increased fidelity.

TELEPHONY

40. Carrier Telephony. Carrier telephony was introduced with a 60 kHz (kilohertz) carrier which could carry 12 simultaneous telephone conversations on one set of wires. Later, higher carrier frequencies were used to permit 48 simultaneous conversations on one set of wires.

41. <u>Type L1 Transmission</u>. The use of coaxial cable for
    long-distance transmission began this year. The ca-
    ble did not have the problems of noise and crosstalk
    which the previous lines and cables were subject to.
    The first type introduced this year was known as
    type L1 with a capacity of 600 channels per pair of
    coaxial cables. One cable was used for each direction
    of transmission. Repeaters were located at approxi-
    mately eight-mile intervals. In June, a coaxial cable
    system was used for two-way television transmission
    between New York and Philadelphia.

TELEVISION

42. <u>Licensing</u>. The Federal Communications Commission or-
    dered a hold on television licensing pending comple-
    tion of a study to determine the best technical stand-
    ards for television.

TIME SIGNALS

43. <u>WWV</u>. On November 6 the Bureau of Standards station
    was destroyed by fire. The station was off the air
    until January 1, 1941.

1941

AMATEUR

44. <u>War</u>. Because of the war in Europe, amateur activity
    had been greatly disrupted. By the beginning of the
    year, complete silencing of the amateur had been or-
    dered in Czechoslovakia, Egypt, France, Lithuania,
    Poland and Switzerland. On December 7, Japan at-
    tacked Pearl Harbor bringing the United States into
    the conflict. General restrictions were put on the
    amateurs with only a few permitted to remain on
    strictly for civil defense purposes. The FCC Order
    # 87 of December 8, 1941, stated:

Whereas a state of war exists between the United States and the Imperial Japanese Government, and the withdrawal from private use of all amateur frequencies is required for the purpose of National Defense;

IT IS ORDERED, that except as may hereafter be specifically authorized by the commission, no person shall engage in any amateur radio operation in the continental United States, its territories, and possessions, and that all frequencies heretofore allocated to amateur radio stations under part 12 of of the Rules on Regulations BE: AND THEY ARE HEREBY WITHDRAWN from use by any person except as may hereafter be authorized by the commission.

By order of the Commission
T.J. Slowie
Secretary

ANTENNAS

45. H. Yagi. The Japanese H. Yagi, inventor of the Yagi beam antenna, attempted to develop the beam for a death ray for use in war. Although it was not successful for this use, it became a very popular form of beam antenna with the amateurs.

BROADCASTING

46. Yankee Network. The first FM chain, the Yankee network, started operation in New England and operated until it was dissolved about 26 years later.

47. FM licenses. Beginning in January and continuing for a few years, the licenses for FM stations indicated the frequencies on which the station operated. At this time, all FM broadcasting was on the band between 42 and 50 MHz. For example, the call W55M indicated a frequency of 45.4 MHz. The M indicated the city, Milwaukee.

48. W9XEN. The power of FM station W9XEN in Chicago was increased to 50 kw, making it the most powerful FM station in the country.

49. FCC Decree. Early in 1941 the Federal Communications
    Commission issued the decree:

> No license shall be issued to a station affiliated
> with a network organization maintaining more than
> one network.

> This action resulted from the concern of the Federal
> Communications Commission about the National Broad-
> casting Company having the Red and Blue networks.
> This was considered as being not in the best interest
> of the public.

COMPONENTS

50. Thermal Relay. A new relay was brought out by Amper-
    ite. The device was used to automatically switch
    equipment from battery power to AC power when the
    AC cord was plugged into the lines. The changeover
    was automatic, with no noticeable break in operation.

51. Constant Voltage Transformer. A new line of voltage
    regulating transformers was put on the market by the
    Sola Electric Company of Chicago. The transformers
    were rated from 25VA to 3KVA. Regulation of the
    secondary output voltage was within 1 percent for line
    voltage variations of 90 to 130 volts.

52. Tubes. In March, developmental models of tubes for
    proximity fuses were shot from guns to demonstrate
    their resistance to shock and suitability for use in
    the missiles.

53. Microtube Laboratories. The Microtube Laboratories in
    Chicago brought out a tube designed for hearing aids
    or similar products which was about the size of a dial
    light. Filament--5/8 volt at 20 to 40 ma (milliamperes).
    Plate current--1 ma or less.

54. T-R Tube. As the crystal detectors in the latest radar
    sets became popular, a means of protecting them
    from the transmitted pulse was required. The trans-
    mit-receive tube (T-R tube) was developed for this
    purpose.

55. **Tubes**. New tubes announced this year included:
General Electric Co.--866A/866; Hytron--HY65, HY67;
Ken Rad--6AH7GT, 12AH7GT; RCA--3Q4, 3S4, 5Y3/
5Y3G, 6SF7, 6SG7, 6SL7, 6SN7GT, 6SS7, 12H6,
12SF7, 12SG7, 12SL7GT, 12SN7GT, 866A/866, 45Z3,
117P7GT, 816, 1631, 1632, 1633, 1634, 8001, 8005,
9001, 9002, 9003; Sylvania--7V7; United--Z225; and
miscellaneous--826, 1625, 1626.

INDUSTRY

56. **Bell Laboratories**. The Murray Hill, New Jersey facility
of the Bell System was opened. The laboratory facilities moved from the West Street building in New
York.

LABORATORIES

57. **RCA Laboratories**. Construction started August 8 on
new RCA Laboratories at Princeton, New Jersey. The
cornerstone was laid November 15. This followed the
incorporation of the research staff as a new department.

NAVIGATION AIDS

58. **Melville Eastham**. Starting this year the Loran System,
suggested by Alfred Loomis in 1940, was developed
into a practical navigation system by Melville Eastham
and his group at the Radiation Laboratory of MIT.
From the completion of tests around September, and
continuing to the end of the war, Loran networks had
been installed to provide hyperbolic grids over half
of the earth or more.

ORDNANCE

59. **Proximity Fuse**. Early in 1941 the Navy Department ad-
vised their contractors to develop an electronic fuse
for antiaircraft use. The proximity fuse was the result.

Tubes for the fuse were required to withstand acceleration forces of about 20,000 times the force of gravity. The first tubes made in this country were handmade.

Attempts at using photocells for the fuse failed. The cells were unable to accept the centrifugal force developed by the shell. An experimental VT fuse was tested at the Dahlgren Proving Grounds by the Bureau of Standards this year and appeared satisfactory.

PHOTO TRANSMISSION

60. <u>Radiophoto</u>. The first radiophoto ever received in New York from Moscow was received July 8 by RCA.

PHYSICS

61. <u>Shoupp, Hill and Stallman</u>. The discovery of the exact amount of energy needed to start fission in atoms of uranium was made by Dr. William F. Shoupp, Dr. Jerold E. Hill, and Dr. F.W. Stallman of Westinghouse.

POWER

62. <u>Wind Power</u>. The first large experimental wind generator system in the United States was the Smith-Putnam machine on a hill near Rutland, Vermont. It operated about 1,100 hours before mechanical failure. It was uneconomical compared to fossil fuels at this time.

63. <u>Grand Coulee Dam</u>. The Grand Coulee Dam generators were put in operation. This dam in Washington State was the largest hydroelectric plant up to this time. Design and construction were overseen by Henry J. Kaiser who had also supervised the Bonneville Dam on the same river. The power capability of the dam was approximately 2,200 MW (megawatts).

RADAR

64. Intercept Receiver. In July, the Office of Scientific Research and Development (OSRD) issued a contract to the General Radio Company for a radar intercept receiver to cover the range of 70-1,030 mc.

65. Model XT-1. In January, the need for a radar system was expressed to the Radiation Laboratory at MIT by the Coast Artillery Corps. By April, the initial experimental equipment was set up on Building 6 at MIT and was tracking within a month. An experimental model, the XT-1, was demonstrated by MIT and tested by the Coast Artillery at Ft. Monroe in February 1942.

66. Plan Position Indicator Radar (PPI). The first microwave radar with PPI presentation used on shipboard was installed in May on the USS Semmes. The system gave about a four-mile range on submarines and eight miles on aircraft.

67 Frederick E. Terman. In December, the National Defense Research Committee, at the request of the Bureau of Aeronautics, was asked to develop a radar countermeasure search receiver and jamming equipment. The project was given to the Radiation Laboratory of MIT under Dr. Frederick E. Terman.

REMOTE CONTROL

68. Aircraft. By August the Navy was conducting experiments to drop depth charges and torpedoes from a radio-controlled plane with a television sight.

TELEMETRY

69. Aircraft. RCA developed a television camera and transmitter weighing about 70 pounds which was used successfully for telemetering in drone aircraft.

TELEPHONE

70. <u>Pulse Code Modulation</u>. A form of pulse code modulation was used during World War II for encrypting voice-modulation communication. It is known as the Sigsally Secrecy System. Sigsally was developed for the US government for maintaining secrecy and is considered the first successful use of pulse code modulation.

TELEVISION

71. <u>Standards</u>. On May 3, the FCC announced that the standards submitted by the National Television System Committee (NTSC) had been officially adopted and commercial television with these standards would be permitted on or after July 1 (21 stations came on at that time). These standards called for a 525 line interlaced 2:1 aspect ratio, 30 fields per second with a 6 MHz bandwidth. The same announcements ordered that data on color TV and recommendations be submitted to permit color TV standardization. In May, the subcommittee on color TV was set up by the FCC.

72. <u>British Standards</u>. In Great Britain, a 405-line standard was adopted this year.

73. <u>French Standards</u>. France adopted a standard of 819 lines for their television transmissions.

74. <u>TV Development</u>. By the end of the year, the development of television was stopped because of the priority of WW II requirements.

TELEVISION (COLOR)

75. <u>J.L. Baird</u>. J.L. Baird demonstrated 600-line color TV using a dual cathode-ray scanning system with a rotating, two-color filter disc. The two sets of images were superimposed on a common screen through an optical system. Alignment was extremely difficult to obtain and maintain.

76. NBC Color TV. RCA and NBC engineers transmitted the first successful color TV broadcast in this country from the Empire State Building.

77. CBS Color TV. CBS made information on its system available to electronic companies and the FCC urged companies to adopt it and begin colorcasting. However, color development was soon stopped because of war projects.

TIME

78. WWV. For about two months the station WWV of the National Bureau of Standards had been off the air because of a fire. On January 1, WWV was back on the air using temporary equipment. The station broadcast on 5 MHz (1 kw) 24 hours a day with time signal. Every five minutes the tone was interrupted for voice announcements. The station was on every day except Sunday from 10 AM to midnight eastern standard time.

TROPOSPHERIC SCATTER

79. André Clavier. André Clavier conducted tests at 3,000 MHz on tropospheric scatter between Toulon, France and ships at sea, and updated information gained by Marconi in 1932.

1942

AIRCRAFT

80. Drone Aircraft. This year, the development of remote-controlled aircraft had gotten to the point of being put into production to permit student gunners to have practice on live targets.

AMATEUR

81. Harry A. Turner. What was considered to have been the world's speed record for sending with a straight hand key was 35 words per minute (175 characters). This was accomplished in November at the US Army Signal Corps School at Camp Crowder, Missouri by Harry A. Turner, W9YZE of Alton, Illinois.

82. Operation Banned. January 9, the FCC issued orders to completely silence amateur radio for the duration of the war.

83. Civil Defense. With the war going on, many amateurs responded to the call for radio operators and joined the military. Some stations were reactivated for civil defense purposes. A second closedown of activities was ordered in January. Over 1,600 reactivations had been authorized through January 1. The reactivations were considered too "loose" and getting out of hand. On January 8, the Defense Communication Board (DCB) asked the FCC to rescind everything to date and Order No. 87-A was issued January 8. It stated:

    Whereas considerations of national defense require complete cessation of all amateur operation; It is ordered, That all special authorizations granted pursuant to order No. 87 by, and they are hereby cancelled.
    By order of the Commission.

COMPONENTS

84. Transistor. As the result of semiconductor studies, the thermistor was developed by Bell Labs. The thermistor became widely used for temperature measurement and control devices in a number of fields.

85. Alternator. The alternator replaced the DC generators on Chrysler automobiles and the change to alternator was being made by other car manufacturers.

86. **Tubes**. Among the tubes released this year were: General Electric--8010R; Hytron--HY1269; Sylvania-- 3LF4, 7W7, 14S7, 1201, 1203, 1204, 1291, 1293, 1457; Western Electric--6AK5; and miscellaneous--2C21, 35Z3, 1005. The production of tubes for proximity fuses reached 500 per day. RCA released the acorn tube for military equipment.

COMPUTERS

87. **Relay Computer**. The Model II relay computer with a tape program input and biquinary error detection came out this year.

88. **Analog Differential Analyzers**. The MIT differential analyzer under construction since 1935 was completed and put to work on military problems. Its completion was officially announced this year. It was the most advanced computer to be built up to this time.

89. **John W. Mauchly**. The idea of a general purpose electronic computer had occurred to John Mauchly, professor of physics at the University of Pennsylvania, Moore School of Electrical Engineering, and he had been considering it for some time. In 1942 he wrote a memo proposing an electronic machine, a possible competitor for a mechanical calculator. The proposal was passed on to the Army Ordnance Department, Ballistics Research Laboratory for consideration.

LABORATORIES

90. **Dr. E.W. Engstrom**. All the research activities of the Radio Corporation of America were consolidated with the opening of the RCA Laboratories in Princeton, New Jersey. Dr. Elmer W. Engstrom was the director of general research.

NAVIGATION AIDS

91. **Loran**. The first Loran stations were put in operation on the coasts of Newfoundland, Labrador, Nova Scotia

and Greenland in October by the Radiation Laboratory of MIT. Fixes could be provided up to 700 miles in the day and up to 1,400 miles at night, with an accuracy comparable to celestial navigation. Later, on the east coast of the United States, Loran coverage was extended from Nova Scotia to Delaware.

In October the first Loran station was put in operation on the western seaboard, operating on about 160 meters. It provided position accurate to within one percent for ships within 300 to 800 NM (nautical miles) and five percent in secondary areas (over 1,000 miles).

92. Robert J. Dippy. The Gee system of radio navigation, similar to the Loran System of the United States, was put in operation around England to permit bomber guidance over Germany. Gee was independently developed in England by Robert J. Dippy of the Telecommunication Research Establishment.

ORDNANCE

93. Information Friend or Foe (IFF). The attack on Pearl Harbor showed the military that it was absolutely vital to have a means of determining whether an aircraft detected by radar was an enemy or a friend. IFF equipment was designed and top priority given to its installation on US and allied aircraft and ships. By the fall, IFF equipment was being installed at the rate of 15,500 sets per month. The following year development of improved equipment was pushed by the National Defense Research Committee.

94. Proximity Fuses. January 29 a pilot production of proximity fuses were tested. Fifty-two percent activated themselves in proximity to water. The Crosley Corp. was authorized to begin a limited production. A second test was run in April during which 70 percent fired. By August, the tests were run from the USS Cleveland against radio-controlled aircraft. The three planes were shot down by only four proximity fuse shells. When mass production started, the shells cost over $700 each.

PHOTO TRANSMISSION

95. **Radiophoto Service.** Radiophoto circuits were opened
    between the United States and Australia by RCA on
    March 20. Other circuits were opened this year to
    Egypt and Europe.

PHYSICS

96. **Dr. S. Benzer.** The physics department of Purdue
    University began a study of germanium this year.
    This study showed how to produce germanium of much
    higher purity and that the electrical properties of the
    germanium is determined by the impurities present.
    Under the direction of Dr. S. Benzer, the high-back
    voltage germanium rectifier was discovered. This was
    a significant step in the development of the transistor.

97. **Enrico Fermi.** Working under the stands at Stagg Field
    of the University of Chicago, Enrico Fermi and his
    associates had collected enough graphite and uranium
    of the necessary purity to form a pile of the size ne-
    cessary to support a sustained nuclear fission reac-
    tion. On December 2, the expected reaction was ob-
    served. Although only about one-half watt of power
    was produced, it was sufficient to show that an atomic
    bomb was possible. This experiment was considered
    to have been the advent of nuclear power.

POWER

98. **Combustion Control.** By this time, some power stations
    were using phototubes to control combustion of fuel
    by monitoring the gas and smoke density in the stack.

RADAR

99. **Aircraft Radar.** Because of the greater effectiveness
    of locating submarines from aircraft radar than sur-
    face ships, a group of ten preproduction microwave
    radar sets were built at the MIT Radiation Laboratory
    and installed on B-18 aircraft. These became part of

the Sea Search Attack Group set up under command
of Colonel William C. Dolan.

100. R.C. Sanders. RCA had developed a low-altitude
radio reflection altimeter under the direction of Mr.
R.C. Sanders. When success was achieved, Sanders
suggested its use for locating targets ahead. It was
called a sniffer when used in this manner. It could
cause a bomb to be dropped at a selected distance
from the target.

101. Luis Alvarez. The development of the Ground-
Controlled Approach (GCA) radar systems was large-
ly the work of Dr. Luis Alvarez and his group at
MIT. The first model made was known as the GCA
Mark I of 1942. This development was completed in
December of 1942. Development began almost imme-
diately on the Mark II.

102. PPI Installation. The first airborn plan-position-
indicator radar was installed in a Lockheed patrol
bomber in July. The radar had been developed at
MIT for submarine detection and tracking.

103. 3-cm Radar. A 3-cm radar system was completed this
year by the radiation laboratory of MIT. This radar
showed a very great improvement in picture detail
over previous models.

104. SCR-582. The Research Construction Co., the model
shop for MIT Radiation Laboratory, built 50 SCR-582
radars for the Signal Corps this year.

RADIO ASTRONOMY

105. J.S. Hey. Reports of radar jamming in February were
investigated by J.S. Hey. Directional checks re-
vealed that the direction from which the signals came
was that of the sun. The noise was caused by sun-
spot activity.

TIME SIGNALS

106. **WWV.** The standard frequency and time service of the Bureau of Standards Station WWV was increased by the addition of 10 and 15 MHz time signals and the musical tone of 440 Hz (hertz). The signals were still coming from the temporary station with a power of one kilowatt.

1943

BROADCASTING

107. **Edward J. Noble.** NBC had been ordered by the FCC to get rid of one network, either the Red or the Blue. The Blue network was sold to Edward J. Noble who wrote a check for $8 million. He then started setting up a new rival of NBC. His network became the American Broadcasting Company (ABC).

108. **FM Call Letters.** Since January 1941, frequency modulated station licenses had been issued in a way to indicate the frequency and location of the station. For a number of reasons, such a system soon proved impractical. The system of all letters was adopted and station W55M became WMFM.

COMPONENTS

109. **Rudolf Kompfner.** The traveling wave tube (TWT) was invented by Rudolf Kompfner, the American physicist. The TWT was a wide-band device for frequencies on the order of 3,500 MHz. It provided bandwidth of roughly 1,600 MHz. The device was developed at Birmingham University in England and by J.R. Pierce of the Bell Laboratories in the United States.

110. **Capacitors.** It was this year that the British started making their own aluminum-coated paper capacitors to reduce component size and weight.

111. Resistors. Resistors having low-temperature coeffi-
     cients and high stability were marketed this year.
     The stability was such that a tolerance of one per-
     cent was maintained for normal operation. They
     were called deposited-carbon resistors after their
     manufacturing process.

112. Tubes. Few new types of tubes were announced this
     year primarily because of the war effort. Tube en-
     gineers in general were improving the existing types
     for longer life and greater reliability for military
     service.
        Many broadcasting stations were in need of tubes
     and were advertising in amateur magazines as a
     source. Amateurs were asked to notify the National
     Association of Broadcasters of any tubes they would
     sell. NAB would then notify the stations in need.
        Sylvania announced a few new tubes, primarily up-
     dated versions of older types, specifically the 1R4/
     1294, 3B7/1291, 3D6/1299, 7C4/1203A, 7E5/1201.
     RCA announced three new cathode ray tubes, the
     3BP1, 3EP1/1806-P, and the 7CP1/1811-P1. Taylor
     brought out the TW75. Among other tubes were the
     2C33, 2C34/RK34, 12A6GT, 12SR7GT and the 1006.

        COMPUTERS

113. Colossus. The British developed the electronic digital
     computer, Colossus, at the Bletchley Park Establish-
     ment. This is considered to be the first electronic
     digital computer in Great Britain.

114. Eniac. After seeing the proposal from John Mauchly
     for an electronic computer written the previous year,
     the army made a contract with the Moore School of
     Electrical Engineering to build an Electrical Numeri-
     cal Integrator and Calculator (Eniac); John Presper
     Eckert Jr. became the chief engineer with Mauchly
     as the consultant. Construction was started the fol-
     lowing year.

INDUSTRY

115. <u>Union Carbide Corporation</u>. Union Carbide entered the nuclear field. Because of its experience in high temperature and pressure operations, it was chosen to design and operate the Oak Ridge Gaseous Diffusion Plant for the separation of uranium 235 from natural ore.

116. <u>Zenith Radio Corporation</u>. The Zenith hearing aid was put on the market by the Zenith Radio Corporation and sold for $40.00. At this time, competitive units were selling for $100 to $200.

LITIGATION

117. <u>Fessenden, Lodge, Pupin, Tesla, Stone</u>. On June 21, the Supreme Court invalidated the patent of Marconi, granted in 1904 on improvements in apparatus for wireless telegraphy by Hertzian waves. It was held that the work of Fessenden, Lodge, Pupin, Tesla and Stone anticipated the claims of Marconi.

ORDNANCE

118. <u>Proximity Fuses</u>. On January 5 the VT fuse (variable time) equipped shells saw action against Japanese aircraft. The USS <u>Helena</u> was the first to use the fuse and brought down a Japanese bomber with the second salvo.

PATENTS

119. <u>G.C. Southworth</u>. G.C. Southworth applied for a patent later granted for a method of artificial vision by scanning terrain in overcast weather for object location, scene ID, etc. by using reflected radio waves emitted by the sun.

PHOTO TRANSMISSION

120. Radio Photo Service. Radio photo service was initiated
      by RCA between New York and Stockholm (February
      22), Dakar (March 10), and Berne (September 21).

PHYSICS

121. Otto Stern. Studies in the quantum effects in atomic
      orientation by molecular beam techniques won the
      Nobel Prize in physics for Otto Stern of the United
      States.

122. Plutonium Production. The first atomic reactor capable
      of producing one million watts of thermal energy was
      developed at the Oak Ridge National Laboratory at
      Oak Ridge, Tennessee. It was put in operation
      November 4 and was known as the "Clinton Pile."
      Its primary purpose was to produce plutonium.

RADAR

123. SCR-584. The first production model of the SCR-584
      radar was delivered in May to Camp Davis. It was
      put in operation at Anzio beachhead in February of
      1944.

124. Radar Development. Because of the war, up to this
      time radar had been developed in secret with little
      information leaked to the public. On April 25 the
      first official announcement was made to the press by
      the War Department. Radar was largely instrumental
      in saving England from the German Blitz in 1940 and
      1941.

RADIO

125. Radio Technical Planning Board. The FCC asked the
      Institute of Radio Engineers and the electronics in-
      dustry to make plans for postwar electronic activities.
      As a result, the Radio Technical Planning Board was
      formed with panels on AM, FM and TV.

TELEPHONE

126. __Undersea Repeaters__. The first submerged cable re-
     peaters were used in a British-installed deep-sea
     cable in the Irish Sea.

TIME

127. __WWV__. The new transmitter of the National Bureau of
     Standards WWV went on the air from Beltsville,
     Maryland. The station used 10 kw with services
     extended to include: standard time intervals syn-
     chronized with the basic time signals, standard audio
     frequencies and the standard musical pitch of 440
     and 4,000 cycles per second. Twenty-four-hour
     service was now available in 5 and 10 MHz with 15
     MHz on in daytime only. Voice announcement of the
     station call letters was given on the hour and half
     hour.

1944

BROADCASTING

128. __American Broadcasting Company__. Initially the name
     American Broadcasting Company belonged to station
     WOL in Washington, DC. In October, the name was
     sold to the Noble interests for $10,000 for use by
     ABC.

COMPONENTS

129. __Relays__. The hermetically sealed contact relays and the
     mercury contact relays came out about this time in
     an effort to provide reliable relays for military equip-
     ment.

130. __Tubes__. New tubes announced this year included:
     Continental--CE-29, Ce-306; Eimac--3C24/24G, 15E,
     25T, 53A, 75TH, 3-25A3(25T), 3-25D3 (3C24), 127A,

152TH, 227A, 304T, 327A, 527; General Electric--
GL-2C44, GL-3C22 (Lighthouse), GL-446A, Gl-446B,
GL-599; Hytron--6AK5, 6A15, 6AQ6; RCA--OA3/
VR75, OC3/VR105, OD3/VR150, 1P28, 2D21, 3B25,
6AK6, 6AL5, 6AQ6, 6J4, 829-B, 9C21, 9C22; and
Taylor--803.

131. Vacuum Switches. A new line of vacuum switches was
      announced by the General Electric Co. The vacuum
      insured dirt-free contact and freedom from oxidation
      corrosion or arcing.

COMPUTERS

132. Eniac. Construction of the Eniac was started by Eck-
      ert and Mauchly and continued into 1945 at the Uni-
      versity of Pennsylvania.

133. Mark I. On August 7 Mark I, the first large-scale
      general-purpose automatic computer was dedicated by
      James B. Conant, President of Harvard, and Thomas
      J. Watson, the founder of IBM. The Mark I was
      built in cooperation with IBM. Although utilizing re-
      lays, the Mark I showed the value of large-scale
      computers and was the predecessor of later high-
      speed electronic computers. This was the machine
      for which Professor Aiken wrote his proposal in
      1937 and was the first automatically sequenced cal-
      culator.

INDUSTRY

134. Allen D. Cardwell. A.D. Cardwell sold the Allen D.
      Cardwell Manufacturing Co. to the Grenby Manufac-
      turing Co. owned by Carl Gray and Ralph Soby.
      The Cardwell Manufacturing Co. was then moved to
      Connecticut.

INSTRUMENTATION

135. John T. Potter. John T. Potter of the Potter Instru-
      ment Company developed a binary decade counter

reported in the June issue of Electronics magazine.
The decade units could be gaged to provide a high
speed-high accuracy pulse counter. The counters
were widely used in digital instruments.

PATENTS

136. Eckert and Mauchly. A patent was filed by Eckert
and Mauchly for the Eniac computer. After considerable discussion, the president of the University of
Pennsylvania agreed that they could file without the
University requiring an assignment of rights. This
action led to much discussion which ended in 1946
with all employees concerned being asked to sign a
patent release form.

PHOTO TRANSMISSION

137. Radio Photo Service. The first radio photo transmission service between New York and Naples, Italy was
started on February 1 by RCA.

PHYSICS

138. Isidor Isaac Rabi. The measurements of magnetic
properties of atomic nuclei by molecular beam techniques brought the Nobel Prize to Isidor Rabi this
year.

139. Y.K. Zavoysky. Soviet physicist Y.K. Zavoysky was
the first to observe electron-spin resonance effects.

PROXIMITY FUSES

140. Proximity Fuse Projectiles. By the end of the year,
87 contractors were delivering parts of the proximity
fuse. The variable-time (VT) fused projectiles were
used against the buzz bombs launched against London. In the last days of the bombings, approximately
79 percent were destroyed in flight as compared to 24
percent with conventional shells.

RADAR

141. <u>AN/APS13 Radar</u>. Developed by the Radiation Laboratory of MIT, the AN/APS13 radar filled the need of a tail warning system for aircraft to warn the pilots of an enemy approaching from the rear. These sets were first used in December on reconnaissance planes in Europe. Danger was indicated by a ringing bell and light indicator in the cockpit.

142. <u>Ground Control Approach</u>. While the first GCA radar was being checked out, the development of an improved model had begun. It was known as the GCA Mark 1. The first preproduction model was ready for testing January 1. It was identified as the AN/MPN-1. The first production units went into operation in Europe, the Pacific area and Alaska in early 1945.

RADAR ASTRONOMY

143. <u>Radar Astronomy</u>. Experiments were started this year to attempt to detect the moon by radar. Attempts using 120 MHz started in the summer of 1944 at the Technical University of Budapest, Hungary. Success was obtained in February of 1946.

144. <u>Grote Reber</u>. Grote Reber and others confirmed Jansky's theory of radio noise origin in the Milky Way and showed that there were powerful noise sources in certain constellations.

RECORDING

145. <u>Recording</u>. When the Allies took Luxembourg in World War II, they found a tape recorder the Germans had developed. It produced the finest recorded sound known up to this time.

TELEVISION

146. <u>J.L. Baird</u>. Baird in England announced his two-color

TV system which had no moving parts but used a dual-beam cathode-ray tube. He was using a mechanical scanning system for pickup at this time.

1945

AIRCRAFT

147. **RF Ignition**. The P.R. Mallory Co. announced a new radio-frequency ignition system for aircraft. The system converted the battery, at the manifold ring, from low voltage to high voltage for the plugs.

148. **Aircraft Radio**. An improved two-way radio for aircraft was produced this year. The system weighed only 15 pounds and permitted use of the radio range system for navigation, communication with control towers, and the reception of broadcast programs.

AMATEUR

149. **Frequencies Released**. Shortly after VJ Day, the FCC released part of the 2-1/2-meter band for amateur use. On November 15 the FCC released the five- and ten-meter bands.

150. **New US Districts for Amateurs**. During the shutdown of amateur activity during WW II, the FCC in cooperation with the American Radio Relay League (ARRL) worked out a new district arrangement more logically in accord with the population and a tenth district was added (∅). Two-letter K prefixes were set up for US possessions with W and K set up for US growth.

BROADCASTING

151. **FM Calls**. By the middle of 1945, the FCC again changed its call system for FM stations. FM stations affiliated with other standard broadcast stations would

use the same call as assigned to the AM station, with the addition of "FM." At this time, WMFM, the Milwaukee Journal Station affiliated with WTMJ, became WTMJ-FM.

152. FM Band. The FCC moved FM broadcasting to a higher frequency band--88-108 MHz to reduce sky-wave interference present on the 42-50 MHz band. This move was made by the FCC in response to a joint request by the Television Broadcasters Association and FM Broadcasters Inc. (June 27).

153. FCC Reports. By this time, frequency modulation had begun receiving wide acceptance. FCC statistics indicate that there were 46 FM stations in operation, seven under construciton and 353 applications for licenses on file.

CIRCUITS

154. Stuart W. Seeley. The ratio detector, a new form of detection for frequency modulation, was disclosed at a meeting of the Institute of Radio Engineers by Stuart W. Seeley on October 3.

COMMUNICATION

155. Taxi Communication. The FCC allocated several frequencies in the 152 to 162 MHz bands for taxicab use. In about two years, over 600 licenses had been issued to companies, with installation in approximately 30,000 cabs.

COMPONENTS

156. Miniature Meters. A line of miniaturized panel meters was put on the market in August. MB Manufacturing Company Inc. produced hermetically sealed 1-1/4-oz. meters, about an inch in diameter.
     Other companies, including DeJur-Amoco, Roller Smith and the Marion Electrical Instrument Co., brought out hermetically sealed 1-1/2-inch instruments.

157. Miniature Transformers. Several new lines of miniature transformers came on the market, one type was announced by the Superior Electric Products Corporation; another line announced by the Permoflux Corporation provided flat response within ± 2 db from 100 Hz to 8,000 Hz.

United Transformer Corporation brought out a line of what it claimed to be the world's smallest transformer. These units were 7/16" x 9/16" x 3/4" in size.

158. Variable Resistors. Miniature variable resistors and potentiometers were put on the market early this year. They were intended for hearing aids or similar equipment. Linear or audio tapers were available.

159. Tubes. The new tubes announced this year by their developers included: Amperex--233; Chatham--3B28; DuMont--Cathode Ray Tubes 5SP, 5RP; Eimac --3X100A11/2339, 4-125A, 4-250A, 3X2500 A3; Electronic Enterprise--EE2BT; Federal--F 5303; General Electric--2C40 (Lighthouse) 2B23; KenRad--6AK5; National Union--NU1Z2; Raytheon--1B48, 2E32, 2E36, 2E42, 2G22, 2C40, 6N4, CK510X, CK-1080, CK-1090; RCA--6AT6, 6BA6, 6BE6, 6CZ4, 12AT6, 12BA6, 12BE6, 35W4, 50B5; Sylvania--CRT's 3AP1, 3BP1, 5AP1, 5BP1, 5CP1, 5HP1; Taylor--TUF-20, TR4OM, 822S; United--893A; and Westinghouse--WL530, WL678.

160. Lock-in Tubes (Loktal). The Sylvania Electric Products announced their line of lock-in tubes designed for rough service. The socket pins were welded directly to the tube elements eliminating a possible source of trouble. A spring in the socket prevented the tube from coming out of the socket due to vibration.

161. J.R. Pierce. The traveling wave tube (TWT), which had been invented some years before by Dr. Rudolph Kompfner, was developed by J.R. Pierce and L.M. Field of the Bell Telephone Laboratories into a practical microwave tube. The tube was improved to provide several kilowatts of power in the C or S bands with a bandwidth of over an octave.

COMPUTERS

162. <u>Model V</u>. The Model V relay computer by Bell Labs
was completed this year. It contained around 9,000
relays and was designed for general purpose application. Two units were produced, one for use at the
Army Ballistic Research Laboratory, and the other
for the National Advisory Committee on Aeronautics.

163. <u>Eniac</u>. The Eniac, designed by J. Presper Eckert and
John W. Mauchly was completed at the University of
Pennsylvania. It was many times faster than the
Mark I, which was presented to Harvard University
in 1944.

164. <u>National Mathematical Laboratory</u>. The British National
Mathematical Laboratory was established in Britain to
coordinate computer work. There were two major
groups working on computers at that time, one at
Manchester University, the other at Cambridge University.

ELECTRONIC HEATING

165. <u>High-Frequency Heating</u>. Electronic means of extending the life of truck tires was utilized at Wright
Field Air Technical Service Command. The high-
frequency heating was used to vulcanize patches on
truck tires. The system heated all parts of the rubber without danger of burning the surface layers.

INDUSTRY

166. <u>Bruce A. Coffin</u>. On May 2 the Hytron Corporation
changed its name to Hytron Radio and Electronics
Corporation with Bruce A. Coffin as president.

167. <u>Masaru Ibuka</u>. The Tokyo Telecommunications Laboratory was established in October by Masaru Ibuka.
The organization eventually grew to the worldwide
organization Sony.

168. <u>Ben Engholm</u>. With the death of Ben Engholm, the

Rola Company was sold to the Muter Company and became a subsidiary of Muter.

LORAN

169. Eastham, Fink, Pierce and Street. The Long-Range Navigation System Loran was developed at the MIT Radiation Laboratory by Donald Fink, J.A. Pierce and J.C. Street under the direction of Melville Eastham.

MAGNETISM

170. Alnico 5. The magnetic material Alnico 5 was being applied this year in speakers and other devices. The high flux density of the material permitted a reduction in the speaker depth and weight with performance as good or better than the electrodynamic types.

MEDICAL

171. Stethoscope. A new electronic stethoscope came out this year using the new miniature tubes available. The device was made by the Maico Company, Inc.

ORGANIZATIONS

172. National Association of Broadcasters. The FM Broadcasters' Association joined with the National Association of Broadcasters to become a department of the National Association of Broadcasters.

173. Joint Electronic Tube Engineering Council (JETEC). JETEC was formed by the Radio Manufacturers Association to coordinate matters concerned with the standardization of electronic tubes. The first chairman was O.W. Pike.

174. Solid-State Research. Federally funded solid-state research had been under way for several years. In

July, at Bell Laboratories, authorization was issued for work on solid-state materials to start with the idea of developing new devices for communication systems. Some months later the group under William Shockley and Stanley Morgan was set up. It was this group that opened up the solid-state era with the development of the transistor.

PHYSICS

175. D.W. Kerst. Electron acceleration to 185,000 miles per second had been accomplished with the betatron, developed by Dr. Donald W. Kerst of Ohio State University.

176. E.M. McMillan and V. Veksler. Both Veksler and McMillan, working independently, noticed that by altering the Dee voltage of a cyclotron, the ions could be kept in phase and with the reduced duty cycle the energy levels of particle accelerators could be increased. These were called frequency-modulated (FM) cyclotrons.

177. Wolfgang Pauli. Wolfgang Pauli, the Austrian physicist, received the Nobel Prize for his studies leading to the Exclusion Principle, stating that no two electrons in an atom can exist in the same quantum state.

POWER

178. HV Transmission. During the 40's, experiments were conducted in Germany in both high voltage AC and DC transmission. Voltages of up to 440 kv DC and 380 kv AC were investigated. The multiconductor system was devised to reduce the corona loss.

179. Linear Motors. Interest in linear motors was given a shot in the arm by a contract with Westinghouse to develop an aircraft catapult system for the Navy called "Electropult."

180. GCA Radar. The first production contract for ground-control approach radars were for Air Force use.

The first units were put in operation in the Pacific, Alaska and Europe. These units were known as the AN/MPN-1 Radar.

RADIO

181. SOS Transmitter. The Radiomarine Corporation of America announced the automatic SOS transmitters for lifeboats April 8. By turning a crank the set sends SOS and direction-finder signals. Range was reported to be about 1,000 miles.

182. Walkie-Talkies. This year, walkie-talkies were made for civilian use and operated in the band of 460 to 470 MHz. The equipment was made by John Meck Industries.

RADIO ASTRONOMY

183. J.S. Hey and G.C. Southworth. The first to document the recognition of the sun as a radio-noise source in England was J.S. Hey, and in this country, G.C. Southworth.

184. R.H. Dicke and R. Beringer. The first measurements of thermal radio emission from the moon were reported by Dicke and Beringer. They used a chopper-type microwave radiometer.

RECORDING

185. Phonograph Records. RCA brought out a new line of nonbreakable high-fidelity phonograph records which they had been developing for over ten years. They were demonstrated to the press in August.

SATELLITE RELAY

186. Arthur C. Clarke. Arthur C. Clarke, scientist and science-fiction writer in Britain, proposed satellite relaying in Wireless World. He pointed out that

at 22,300 miles high, the satellite would appear stationary and would cover most of the earth's surface except for small regions around the poles. He also proposed manned space broadcasting.

TELEPHONE

187. Pulse Time Modulation. Telephone circuits using pulse time modulation permitting 24 two-way conversations on one 1,300 MHz channel, were set up on an experimental basis over an 80-mile path from the ITT building in Manhattan to Telegraph Hill near Hazlet, New Jersey and also to Nutley, New Jersey.

TELEVISION

188. Projection TV. RCA brought out its new large-screen TV (16" x 21") utilizing automatic frequency control and a relatively small (five-inch) high-voltage cathode-ray tube (CRT). The optical system provided a picture approximately five times the tube screen size.

189. DuMont Television. DuMont Laboratories demonstrated projection TV with a 3' x 4' image.

190. Albert Rose, Paul K. Weimer and Harold B. Law. The image orthicon, invented by Rose, Weimer and Law, was put in service to replace the iconoscope in RCA TV cameras on October 25. The new tube greatly aided the pickup of live programming because of its increased sensitivity.

191. 3-D Color TV. RCA demonstrated field sequential color TV and 3-D color to industry.

192. Harry S Truman. The first time a United States president was seen on a television network is reported to have been October 27, when Harry Truman was seen on television at the Navy Day celebration in Central Park, New York City.

193. Stratovision. About the middle of the year a plan to use airborne stations to serve as relay stations for

television and FM programs was proposed by Westinghouse. This provided wide area coverage with no
wire networks required. The system was termed
"Stratovision."

TIME

194. <u>Isidor Isaac Rabi</u>. The suggestion of using molecular
transitions in an atomic clock is attributed to Isidor
Isaac Rabi, who made the suggestion in a lecture
before the American Physical Society in January.

TRANSISTOR

195. <u>Solid-State Physics</u>. At the Bell Telephone Laboratories, a project was set up to study and relate the
known facts of solid-state physics to the anticipated
needs of the communications technology. This work
culminated in the development of the first transistor.
The junction transistor was proposed by W. Shockley in a patent application this year. The grown
junction transistor was produced several years later.

TRANSPORTATION

196. <u>Railroad Radio Service</u>. Railroad radio service was authorized by the FCC in December. This service
covered the communication from caboose to engine
or from one train to another, terminal to train, and
general yard operations, including remote control.

1946

AIRCRAFT

197. <u>Remote Control</u>. In January, it was reported that two
B-17 flying fortresses had flown from Hawaii to Muroc
Army Air Field in California the previous August,
both without pilots. The drones were controlled from

a third plane which controlled all flight operations
by radio. This test included a simulated bomb
drop-off on Santa Rosa Island.

AMATEUR

198. Dr. A.H. Sharbough and R.L. Watters. On May 18,
two-way voice communication was established over
800 feet on 21,900 MHz. Dr. A.H. Sharbough
(W1NVL/2) and R.L. Watters (W9SAD/2) of the General Electric Research Laboratory, Schenectady, New
York operated two-way in the highest frequency
band allotted to amateurs.

199. Amateur Bands On Again. The bands were returned
to the use of the amateurs sporadically during the
year. By the end of November, all bands had been
returned and were in use.

200. Amateur DX Records. A number of new distance records were made this year by amateurs in the VHF
and higher bands with CE1AH and J9AAO making the
record for distance over a 10,500-mile path on 2,300
MHz. W6IFE/6 worked with W6ET/6 over 150 miles
on 3,300 MHz.

ATOMIC ENERGY

201. Atomic Energy Act of 1946. The Atomic Energy Act
established two governmental groups to control the
development of atomic power or other uses of the
atom in this country. The Atomic Energy Commission (AEC) was the administrative branch with the
Joint Committee on Atomic Energy (JCAE) serving as
the legislative branch. The first chairman of the
AEC was David Lilienthal, Administrator of the Ten-
nessee Valley Authority.

BROADCASTING

202. Stereophonic Broadcasting. What may have been the
first instance of stereophonic broadcasting occurred

in the Netherlands. It was heard in Great Britain, France and elsewhere.

COMPONENTS

203. <u>Dr. Robert Adler</u>. The phasitron, a new tube developed by Dr. Robert Adler of the Zenith Radio Corporation for FM phase-modulated transmitters, was described by Dr. Adler at the meeting of the Chicago section of the Institute of Radio Engineers (IRE) at the February 15 meeting.

204. <u>Diodes</u>. Sylvania announced development of the 1N34 germanium crystal diode in February.

205. <u>Capacitors</u>. Cornell-Dubilier Electric Corporation announced a line of midget capacitors type ZN, for such devices as hearing aids and pocket radios or other limited-space uses.

206. <u>Phono Pickup</u>. The National Co. announced a new velocity type phonograph pickup in March. The frequency response was uniform within 1 db, to about 15 Hz.

207. <u>Speakers</u>. Altec Lansing Corporation introduced the Model 600. The speaker reproduced the low and high frequencies on different cones (Dia-Cone Principle) giving excellent reproduction over the audio range.

208. <u>Transformers</u>. The United Transformer Corporation announced its line of midget transformers. The so-called sub-ouncers weighed only about one-third of an ounce. This was a major step in the miniaturizing of electronic devices.

209. <u>Dr. Ernest O. Lawrence</u>. A color television tube known as the Chromatron was invented by Dr. Ernest O. Lawrence of the University of California. This was one of the early forms of a color tube; however, it was not completely satisfactory for use at this time. Work on improving the tube continued for some years.

210. Projection Kinescope. A new picture tube for projection television systems was disclosed by RCA at the January 24 meeting of the IRE. The tube gave approximately 50 percent more light efficiency than the previous tubes.

211. Vibrotron. The Radio Corporation of America announced the development of the Vibrotron tube. This device converts mechanical motion to a variable electron flow. It was a developmental model available to those having a reason to experiment with it.

212. Tubes. Some of the tubes marketed this year were: Aireon--AT340, 3D26, 4C32; DuMont--7EP4 (CRT); Eimac--4X150A, 4X500A, 35TG; Electronics Products --207M, 891, 892M; General Electric--2C39, 2C43, GL5C24, 7C29, GL502A, GL7D21, GL592, GL5D24; Hytron--2E25, 2E30, HD59; Lewis--3C28, 3D23, 4C32, 4C36, AT257-C, AT340; Machelett--889A Improved type; Raytheon--2E31, 2E35, 2E41, 2G21, RK-4D22, RK-4D32, RK-6D22, Ck505AX; RCA-- 2C43; Sylvania--1LG5, 4C35, 7AG7, R1130, 1B35, 1B37, R4330 (flash); Taylor--TB-35, 833A (improved); TungSol--117Z3; United--5562 (United also announced a line of graphite plate tubes including HV-18, KU23, V-70-D, 812H, 838, 849, 949H); and Westinghouse--WL473.

COMPUTERS

213. George A. Philbrick. George A. Philbrick brought out his operational amplifiers. These were the first DC amplifiers specifically designed for analog computer use.

214. Eniac. The era of electronic computers has been considered by some to have started this year with the development of the Eniac at the University of Pennsylvania. This computer utilized approximately 18,000 vacuum tubes. It was completed in February at a cost of over $486,000. Approximately 150 kw of power was required for its operation. This computer could operate about 1,000 times faster than the Mark I Computer. Although demonstrating the high-

speed calculation possibilities (over 5,000 calcula-
tions/sec.), the programming was complicated and
required hours for a simple program change.

215.   **F.C. Williams.** F.C. Williams came to Manchester University to perfect his computer storage tube and
start work on a stored-program computer. The Williams tube indicated a 1 or 0 by writing a dash or
dot on the tube. A metal plate on the face of the
tube formed a capacitor. Sweeps of the electron
beam could sense the presence of the dots or dashes.
The same beam could regenerate the stored bits.

216.   **Dr. John W. Mauchly and J.P. Eckert.** Work was
started on the construction of the Electronic Discrete
Variable Automatic Computer (Edvac). The computer
had been designed by Dr. John W. Mauchly and
J.P. Eckert of the University of Pennsylvania staff.
The computer was built by University students and
staff members.

217.   **Electronics Control Co.** After the Eniac had been
dedicated, the University of Pennsylvania employees
were asked to sign a patent release form. Eckert,
Mauchly, and some others resigned rather than sign
the release. A short time later the Electronics Control Company was formed by Eckert and Mauchly in
order to complete the study contract awarded them
by the National Bureau of Standards. This was a
study of the computer required by the Census Bureau and a scale model of two mercury delay tubes
for a storage device. The Electronics Control Company later became the Eckert-Mauchly Computer Corporation.

218.   **Maurice V. Wilkes.** The British-designed Electronic
Delay Storage Automatic Calculator (Edsac), built on
the principles of von Neumann, was started by Dr.
Maurice V. Wilkes at the end of the year. This is
the first stored-program computer to be completed.
The computer used 32 internal delay lines giving a
capacity of 1,024 words. Edsac was completed in
1949.

219.   **John von Neumann.** Mathematician John von Neumann

proposed a computer with a memory capable of storing programs as well as the other necessary information for the calculation. The idea was soon universally adopted.

ELECTRONIC MUSIC

220. Electronic Guitar. The electric guitar was developed by the Fender Guitar Co. about this time.

INDUSTRY

221. Tektronix. Tektronix Inc. was founded on January 2, by Howard Vollum and Jack Murdock. Their first development was the 511 oscilloscope. This was a calibrated oscilloscope with automatic triggering capability to 10 MHz.

222. Electronic Valves, Ltd. Electronic Valves, Ltd. was organized by A.C. Cossor Ltd., of London and Sylvania Electric Products Inc. to be a manufacturer of Sylvania tubes in Great Britain.

223. Iskra. Iskra Electrotechnics and Mechanics Factory was founded March 8. In about 30 years it became the leading manufacturer of electronic and electrical equipment in Yugoslavia and gained worldwide recognition.

224. Radio Corporation of America (RCA). A new plant was opened in Lancaster, Pennsylvania to produce larger kinescope tubes for television.

225. M. Ibuka and A. Morita. In May the Tokyo Telecommunications Engineering Co. Ltd. (formerly the Tokyo Telecommunications Laboratory) was incorporated by Masaru Ibuka and Akio Morita.

INSTRUMENTATION

226. Sniperscope. Use of the sniperscope in the Pacific theatre was disclosed by the Army in April. The

sniperscope was developed by the RCA Laboratories and was an infrared image tube permitting night vision of objects illuminated by infrared rays.

227. Signal Generators. Pulse generators and square wave generators appeared on the market with the Hewlett-Packard Model 212 being one of the first. It provided 5 kHz pulses having a 20 NS rise time.

NAVIGATION SYSTEMS

228. Decca. The Decca system was brought out in England and operated on 75-80 kHz. The system utilized phase comparison to determine position. The accuracy capability was about 150 yards at 250 miles in the day, varying to about 800 yards at night.

229. Shoran. In January RCA disclosed its Shoran System as an aid to pilots in blind bombing operations. It claimed to have many peacetime applications in the mapping of previously uncharted areas of the world.

PATENTS

230. M.F. Jones. On July 30, M.F. Jones was awarded US Patent No. 2,404,984 for a three-phase linear motor catapult system.

PHYSICS

231. Mesons. Physicists of the General Electric Research Laboratory are reported to have succeeded in artificially producing mesons for the first time using the 100,000,000-v betatron. The meson is one of the chief constituents of cosmic rays.

232. Bloch, Hansen and Packard. Felix Bloch, William Hansen and Martin Packard were the first to observe nuclear magnetic resonance in this country.

POWER

233. Nationalization of Electric Power. Between 500 and
     600 privately owned power companies in Great Britain
     were taken over and nationalized under the British
     Electricity Authority.

234. US Rubber Co. Late in the year, the US Rubber Com-
     pany was substituting aluminum wire for copper in
     making wire and cables. This was due to the cop-
     per shortage.

RADAR

235. Ground Control Approach. Pan American World Air-
     ways built the first ground-controlled approach sys-
     tem in December at Gandar, Newfoundland. The idea
     was apparently originated by Luis Alvarez, a nuclear
     scientist.

RADIO

236. Pocket Ear. The so-called "pocket ear" was developed
     by NBC and consisted of a pocket-sized unit with an
     earplug. It provided communication between the
     studio stage and the control room without the trailing
     wires of conventional systems.

RADIO ASTRONOMY

237. Project Diana. Project Diana had been authorized by
     a Pentagon directive to study a possible means of
     detecting ballistic missiles that might be launched
     against the United States. The project was under
     the direction of Lt. Col. John De Witt, Jr. Almost
     immediately after VJ day a group including E.K.
     Slodola, Dr. Harold Webb, Herbert Kauffman and
     Jacob Mofenson, radar engineers of the Evans Signal
     Corps Engineering Laboratory began work on the
     project. On January 10, Harold Webb and Herbert
     Kauffman of the US Army Signal Corps received a
     0.25-second pulse reflected from the moon. This was

on 110 MHz and the first proof that electromagnetic waves could penetrate the ionosphere. The first man credited with hearing the echo was Herbert Kauffman, W2OQU. After repeatedly seeing the signal on the oscilloscope there could be no doubt; some weeks later the achievement was made public.

238. Zoltan Bay. Almost simultaneously, Zoltan Bay in Hungary, who had proposed the moon experiment several years earlier, received a lunar return. The announcements of both were made the same month.

RAILROAD RADIO

239. Railroad Communication Tests. In July of 1946, tests were conducted for the railroad. These radiotelephone communication tests were held on the main line of the Nickel Plate Road for the Association of American Railroads. Tests were to permit a study of the performance of VHF equipment on high-speed freight trains for end-to-end and train-to-train service

240. Railroad Music. The Atchison, Topeka and Santa Fe Railroads announced the installation of equipment on railway cars for providing recorded music in roomettes, drawing rooms and dining rooms. The system was designed by Farnsworth TV and Radio Corp.

RECORDING

241. High-Fidelity Recordings. London Records announced a new disk having a frequency response of from 40 to 14,000 Hz flat to ± 1 db.

STANDARDS

242. International Committee on Weights and Measures. In October, the International Committee on Weights and Measures met with delegates from the National Bureau of Standards and similar organizations from countries around the world. Delegates met in Paris to consider the units of light and electricity presently in use as

measurement requirements continued to require more and more accuracy. It became apparent that there was a difference between measurements in international units based on the cgs system and the absolute system derived from fundamental mechanical units of mass, length and time. The Committee agreed that the absolute units would supersede the international units.

The change which took effect January 1, 1948, made the following adjustments in the US:

1 international ohm   = 1.000495 absolute ohms
1 international volt   = 1.00033 absolute volt
1 international ampere = 0.999835 absolute amp
1 international henry   = 1.000495 absolute henries

TELEPHONE

243. Coaxial Cable. Regular coaxial cable relay service was inaugurated between Washington DC and New York City, principally for TV use.

244. Mobile Service. The first commercial mobile telephone service was started in St. Louis, Missouri on June 26.

245. Microwave Service. The first microwave system for telephone use was installed for communication between Los Angeles and Catalina Island. The system provided eight two-way channels.

TELEVISION

246. Richard Thomas. Richard Thomas developed a color system using a three-lens camera for developing the three-color images. The system was originally intended for color pictures but was not acceptable for TV.

247. Allen B. DuMont. About this time, Allen B. DuMont started the DuMont Television Network with station WABD and W3XWT (later WTTG) in Washington, DC. The network never gained wide acceptance and after some years was sold to Metropolitan Broadcasting Co.

248. **TV Receivers**. RCA Victor Division put on the market
the first postwar TV sets. The basic model was the
630TS, a 10" tube set. This is believed to be the
first mass-produced and -marketed TV set.

249. **Color TV**. CBS demonstrated its color system for the
press and representatives of the industry. The
transmitter was built by Federal Telephone and Radio
Corp. and operated on 490 MHz with 1 kw power to
a 20-db gain antenna.

1947

AMATEUR

250. **Quarter Century Wireless Association, Inc. (QCWA)**.
The Quarter Century Wireless Association was
planned on November 4 by W2FX, John Di Blasi;
W2UD, Uda Ross; W2RF, John Gioe; W2EX, Ed Crane;
W2DX, Irving Groves; and W2DI, Dr. Ernest Cyreax.
The occasion was a ten-meter roundtable. The first
meeting was held December 5 in New York City.
Thirty-four charter members were present. The
purpose of the organization was to develop friend-
ships and cooperation among amateurs having 25 or
more years' standing.

251. **SSB**. The real start in single sideband operation came
this year when several amateurs began working on
single sideband and showed that it could be copied
on a normal communication receiver. From that time,
sideband operation grew to eliminate the use of amplitude modulation by most amateurs.

ARC WELDING

252. **Rod Production**. By 1947 the electrical welding industry had grown to the point of requiring over 350
million pounds of welding rods. This showed a
growth of over 1,400 percent in the use of the rods
in just 15 years.

BATTERIES

253. New Dry Cells. The new Eveready high-energy cells were introduced by the National Carbon Co. The new cells carried double the energy of the previous cells of the same size. The improvement was made by a change in chemicals used in developing of the VT fuse cells.

CITIZENS BAND

254. CB Authorization. Citizens band (CB) operation was first authorized in 1947. CB refers to the Class D Citizens Radio Service.

COMPONENTS

255. Relays. The Stevens Arnold Co. announced a new relay capable of 1 ms (millisecond) or less operation with a contact rating of 0.5 amp at 110 v. The relay was about the size of the average metal tube.

256. Wire. Possibly the finest wire ever produced was made by the Westinghouse Lamp Division for the Bell Telephone Laboratories. The tungsten wire was 18 hundred-thousandths of an inch in diameter. The wire was made for electric lamp and tube filaments.

257. Capacitors. The Solar Mfg. Co. announced its new line of miniature electrolytic capacitors (Type LB). This component permitted great size reduction for electronic equipment.

258. Cornell-Dubilier. A new line of flat midget capacitors was announced by Cornell-Dubilier Electric Corporation. The ZN line was designed specifically for hearing aids and pocket radios. Capacity ranged from 0.0001 mfd. to 0.1 mfd., and the capacitors were rated from 150 to 600 v.

259. Crystal Diode. A new crystal diode, the IN38 germanium rectifier, was announced by Sylvania. The peak inverse voltage was 100 v at 22 ma.

260. <u>Teflon</u>. The new plastic, Teflon, was being applied to capacitors and other components requiring high insulating properties at high frequencies and temperatures. It could withstand all known solvents and temperatures to 575 degrees C. and had essentially zero water absorption.

261. <u>Tubes</u>. The introduction of new tubes to the industry continued this year. Some of the companies and their products were: Amperex--AV3K, AV3E; Chatham--5594, 1Y2; Eimac--4-400A, 4-500A, 4-750A, 4-65A; Federal--9C28, 9C29, 9C30, 9C31, 5563; General Electric--3Y125000A3, GL4D21, 4SN1A1, 6T8, 10FP4, 12AT7, 12AU7, 19T8, GL-5544, 5545, 5630, 5648, 5663, 5665, 5691, 5692, 5693, 5674; Heintz & Kaufman--HK27, HK57, HK257C, HK357C; Hytron--1X2, 3B4, 6BQ6, 25BQ6, HY75A, 5516; National Elex--NL710, NL714; Raytheon--CK605CX, CK608CX, RK61; RCA--OA5, 1P37, 1P42, 1U5, 2E26, 2X26, 3D24, 12AX7, 12AL5, 6BJ6, 1945, 1946, 1947, 1950, 5527, 5653; and Sylvania--OA5, 3D24, 7F8, 14F8, GH302, GG304, R4300, R4340.

262. <u>Vacuum Gauge</u>. RCA introduced a new line of vacuum gauge tubes utilizing a palladium plate which is porous to hydrogen when hot, but a barrier to all other gases. Five types of the gauge were produced.

COMPUTERS

263. <u>Mark II</u>. The Mark II version of Harvard's Sequence-Controlled Relay Calculator was constructed at the Harvard Computer Laboratory, and tested from July to January 1948. It was set up that year at the Navy Proving Ground, Dahlgren, Virginia. This computer used about 13,000 relays operating in about 1/100 second; the speed was considerably faster than that of the Mark I version.

264. <u>Berks, Goldstine, von Neumann</u>. The historic document "Preliminary Discussion of the Logical Design of an Electronic Computing Instrument"--published this year by Berks, Goldstine, and John von Neumann--was considered the basis for the design of a number of electronic computers.

265. Binac. Because of delays in getting started on the Census Computer, and to get money to carry on, Eckert and Mauchly of the Electronic Control Co. signed a contract with the Northrop Aircraft Co. to design and build a binary automatic computer (Binac) which was completed in 1949 and was the first computer built in the United States which included an internally stored program.

266-7. Edsac. The Electronic Delay Storage Automatic Calculator (Edsac) was started in 1946 at Cambridge University. Work continued for about three years before completion. The machine's memory was composed of 128 thermostatically controlled acoustic delay lines, each capable of storing 384 bits of information as sound waves in mercury. The access time was 48 to 384 microseconds (μs).

CONFERENCES

268. Seventh International Radio Conference. The seventh International Radio Conference convened on May 16 in Atlantic City, NJ, specifically to revise the 1938 Cairo regulations where necessary to meet modern conditions and services. Seventy-nine countries attended. Many allocation problems were considered because of new fields such as radar, radio aids to navigation and international aviation. General frequency assignments were increased up to 10,000 MHz. In October, the conference was closed.

269. International Meetings on Marine Radio Aids to Navigation (IMMRAN). In May, 22 nations met in London to set up rules and general specifications for the use of radio and other navigational aids. The capabilities of radio were discussed and recommendations were made for frequency allocations. It was unanimously agreed upon that centimeter radar was practically indispensable, however pulse duration was recognized as a problem. In order to detect objects at short ranges in restricted waters, a pulse length of not over 0.2 microseconds and preferably less would be required.

270. International Telecommunications Conference. Starting

on July 2 and running concurrently with the seventh
International Radio Conference, the International
Telecommunications Conference met to revise the
Madrid Convention of 1932. Many of the same dele-
gates were involved in both conferences.

271. International High-Frequency Broadcasting Conference.
In August, the International High-Frequency Broadcasting Conference convened and further delayed
progress of the Radio Conference (above). Because
of the other conferences this one became a preliminary conference to plan further conferences to consider the orderly placing of the world's broadcasting
stations for effective use of broadcasting allocations.

FABRICATION

272. A.W. Franklin. The idea of stamped wiring was considered and developed for the production of five-
tube radio sets. The idea was originated by A.W.
Franklin of Franklin Air Loop Corp. It never received wide acceptance, undoubtedly because of the
development of the printed circuit technique of the
Bureau of Standards at about the same time.

HOLOGRAPHY

273. Dennis Gabor. The idea of the hologram occurred to
Dennis Gabor on Easter. Gabor, of the London Imperial College of Science and Technology, invented
holography in an attempt to improve the resolution
of the electron microscope. Using coherent light,
he was sure he could improve the image. Results
were encouraging. At that time little practical use
was found for holography and development had
stopped by 1955 because of the difficulty of obtaining coherent light. Interest was revived later with
the development of the laser.

INDUSTRY

274. Motorola. The Galvin Radio Co. changed its name to
the Motorola Corporation.

275. Microwave Heating. Microwave heating came into use
     by industry about this time. The first applications
     were fabric drying and rubber curing. The same
     year General Electric announced an electronic oven
     for heating frozen foods.

276. Joseph Resnick. Channel Master Inc. was organized
     by Joseph Resnick to manufacture television anten-
     nas. Starting with $7,000 of borrowed money, in
     20 years he had built the company to where it was
     sold to the Avnet Conglomerate for $50 million.

277. Eckert-Mauchly Computer Corporation. In order to be
     able to sell stock to raise finances to continue oper-
     ations, Eckert and Mauchly incorporated as the
     Eckert-Mauchly Computer Corporation thus dissolving
     the Electronic Control Co. John W. Mauchly became
     president of the organization with J. Presper Eckert
     Jr. as vice president and chief engineer.

     PHYSICS

278. Atom Smasher. The General Electric Research Labora-
     tory has produced a beam of 70 million-volt X rays
     from its synchrotron, a compact atom smasher.

279. Sir Edward Appleton. Sir Edward Appleton of England
     was awarded the Nobel Prize in physics for the dis-
     covery of the layer which reflects radio short waves
     in the ionosphere.

280. E.M. McMillan. The first electron synchrotron to be
     built was completed under the direction of E.M. Mc-
     Millan at the University of California. Electrons
     were accelerated to energy levels of about 300 MeV
     (million electron volts). It was reported that a par-
     allel development in the Soviet Union resulted in en-
     ergy levels of 500 to 1,000 MeV.

281. Post Roads Act. The Post Roads Act was passed in
     1866 and authorized the Postmaster General to fix
     the rates annually for government telegrams. In
     1947 the Post Roads Act was repealed and no more
     special domestic telegraph rates were specified by
     the federal government.

RADAR

282. "Chirp Radar." The Bell Telephone Laboratories developed the "chirp radar." With this system, long
pulses were transmitted and compressed in the receiver. The system provided both long-range and
high resolution.

283. Speed Meter. Radar techniques were applied to the
development of a speed meter. Speed ranges of
from 0-100 mph were indicated to within 2 mph.
The meter was designed by engineers of the Automatic Signal Division of Eastern Industries, Inc.
Some years later, it was widely used by the police
in many areas.

284. 3.2 Centimeter Radar. The Radio Marine Corporation
of America demonstrated a 3.2 cm radar aboard the
USS American Mariner for delegates of an international meeting on Radio Aids to Navigation. The
radar was capable of detecting objects as close as
80 yards as well as clouds or other objects at 30
miles.

RADIOS

285. Clock Radios. It was around the end of the year that
table radios came out containing alarm clocks. One,
if not the first, radio with a clock was introduced
by the Garod Radio Corporation.

TELEPHONE

286. Mobile Telephone. A mobile radio telephone service
was started by the Bell System this year in the
larger cities for both urban and highway service.
The urban system utilized frequencies in the 152-162
MHz band. The highway service operated in the 30-
44 MHz band.

287. Microwave Telephone System. An experimental microwave system built by the Bell System linked New
York and Boston and operated between 3,900 and

4,200 MHz. This was known as the TDX system.
The system could handle around 240 voice channels.
The success of this system led to the improved TD2
system of 1950.

TELEVISION

288. Zoomar Lens. On July 21, the Zoomar Lens was the
      first of its kind to be used on TV by WCBS-TV,
      eliminating the triturret assembly. The baseball
      game between the Brooklyn Dodgers and the Cincin-
      nati Reds was covered with a smooth continuity nev-
      er before achieved.

289. Color TV Demonstrations. All-electronic, large-screen
      color television was demonstrated on April 30 at the
      Franklin Institute in Philadelphia. The picture was
      8' x 10'. A number of color TV demonstrations were
      held during 1947 and 1948. Three kinescopes were
      used with the composite optically produced.

290. Networks. TV networks began to expand this year
      with a cable from New York to Boston. One net-
      work expanded from New York to Baltimore and
      Washington. Another from Philadelphia, Schenec-
      tady and Washington.

291. RCA. RCA offered to disclose complete details of its
      color system to other manufacturers, a step toward
      acceptance of the compatible television system used
      by RCA.

292. President Truman. The first telecast from the White
      House was reported to have been made by President
      Truman on October 13; this was a request to the
      nation for food conservation.

TIME SIGNALS

293. WWV. Radio Station WWV of the Bureau of Standards
      now operated 24 hours a day and had added frequen-
      cies of 20, 25, 30 and 35 MHz in addition to the
      previously used frequencies of 2.5, 5, 10 and 15

MHz. Signal accuracy on 440 and 4,000 CPS was
one part in 50 million at the source.

TRANSISTORS

294. <u>Bardeen and Brattain</u>. On December 23, Dr. John
Bardeen and Dr. W.H. Brattain made the first practical transistor which consisted of two point contacts
on a slab of germanium crystal. On December 24,
Dr. Brattain recorded in his notebook: "This circuit was actually spoken over and by switching the
device in and out, a distinct gain in speech level
could be heard and seen on the scope presentation
with no noticeable change in quality."
The era of the transistor (so named by Dr. John
R. Pierce) had begun.

ULTRAFAX

295. <u>Ultrafax</u>. On October 21, the Radio Corporation of
America and Eastman Kodak Co. gave the first public demonstration of a high-speed form of information
transfer at the Library of Congress, Washington,
DC. The system was experimental, combining photographic, facsimile, and television techniques. The
goal of the developers was one million words a minute.

1948

AMATEUR

296. <u>Single Sideband</u>. The advantages of single sideband
suppressed carrier (SSSC) had been recognized by
this time. Several stations were on the air with
more under construction. SSSC offered many advantages over amplitude modulation in that it required only one-half or less the bandwidth, without
the heterodyne whistles previously present, and
more efficient use of the station's power. This represented the beginning of a revolution that was soon

to practically eliminate the amplitude modulated stations from the amateur bands.

The first SSSC transmissions were put on the air in September by W6YX and WØTQK on the 14 MHz band.

BROADCASTING

297. Low-Power FM. The FCC authorized low-power (10 watts) FM broadcasting on educational FM channels. This permitted schools to provide FM over a limited area at a low cost.

CHAIN BROADCASTING

298. Mutual Broadcasting System. By this time the Mutual Broadcasting System had over 500 affiliated stations in its chain and advertised itself as the world's largest network. In general the stations were of lower power than those of the ABC or NBC networks.

299. Liberty Broadcasting System. The Liberty Broadcasting System was started as a baseball network in Texas. Starting with KLIF in Dallas and serving about 40 stations, it grew to almost 240 stations by 1950.

CIRCUITS

300. James Clapp. James Clapp, of the General Radio Co., developed the "Clapp Oscillator," a series-tuned oscillator of unusual stability and wide tuning range.

COMPONENTS

301. Transistors. On July 1 the New York Times announced the transistor: "A device called a transistor, which has several applications in radio where a vacuum tube ordinarily is employed, was demonstrated for the first time yesterday at the Bell Telephone Laboratories."

302. <u>Tubes</u>. Some of the new tubes released this year included: Amperex--492R; Bendix--TT-1; Eimac--4-
150A, 4X150G; General Electric--6AV6, 12AV6; Mach-
lett--ML-5658; Norelco--(CRT) 3QP1; Raytheon--
6SA7GT, 6SJ7GT, 6SK7GT, 6SQ7GT, 12SA7GT,
12SJ7GT, 12SK7GT, 12SQ7GT, Ck-5703/CK608CX,
CK571AX, CK5704/CK606BX, CK5702/CK605CX; RCA
--2K26, 5WP15, 6AR5, 6AS5, 6AV6, 6BA7, 6BH6,
12AV6, 12BA7, 19J6, 3SC5, 50C5, 672A, 5618, 5651,
5652, 5691, 5692, 5693, 5696, 5713; and others--
6W4GT, NL617, TGC-1, TGC-2.

303. <u>General Electric</u>. A new magnetron had been developed
at the General Electric Research Laboratory which
was capable of developing 50 kw output at 1,000
MHz. It was developed under a Signal Corps Laboratory contract. This was the highest CW (continuous wave) power ever generated at this frequency.

304. <u>Metal Picture Tubes</u>. RCA announced the development
of the first metal picture tube. This was the round
16-inch tube.

COMPUTERS

305. <u>Selective-Sequence Electronic Calculator</u>. This computer was tested for some time before being announced on January 27. It was designed by IBM
to solve scientific problems. The computer used
about 12,500 tubes and 21,500 relays. It was installed in 1947 in the IBM offices in New York City.

306. <u>Eckert and Mauchly</u>. After many reviews and delays,
the report submitted by Eckert and Mauchly was accepted in June on a fixed-fee agreement for a high-
speed electronic computer with an acoustic delay line
memory and magnetic tape storage. It was later
known as the Univac. By August an agreement was
made with the American Totalisator Co. of Baltimore,
which agreed to invest about a half million dollars
over the next two years for 40 percent of the voting
common stock in the Eckert-Mauchly Computer Corp.
This provided the money to carry on.

ELEVATOR CONTROL

307. Elevator Control. The Otis Elevator Co. developed an electronic scheduling system to provide greater speed and efficiency for elevators. Essentially it measured the waiting time of the passengers and dispatched an elevator to the proper floor after a predetermined time.

FREQUENCY STANDARD

308. Hershberger and Norton. A highly accurate frequency standard based on the effects of radio on certain gases was described by W.D. Hershberger and L.E. Norton in RCA Review (March 1948).

INDUSTRY

309. Muter Co. The Muter Company acquired the assets of Jensen, which became a wholly owned subsidiary of Muter.

INFORMATION THEORY

310. Dr. Claude Shannon. The concept of information theory was discovered by Dr. Claude E. Shannon, an engineer of Bell Labs. He described his concept in his paper, "A Mathematical Theory of Communication" published in the (June/July) 1948 Bell System Technical Journal. This has become the classic work on information theory. By this concept, the ultimate communication capabilities of a specific system could be predicted. This work has also been credited with stimulating the development of Pulse Code Modulation (PCM).

INSTRUMENTATION

311. Magnetic Recording. By this year, magnetic recording had been improved to where it was indistinguishable from the original sound. The Bing Crosby Show was

the first show to go on the air directly from the original tape (May 1948).

312. Columbia Long-Playing Records. In June, Columbia Records announced the new long-playing records (LPs) operating on 33-1/3 rpm (revolutions per minute). The discs permitted up to 50 minutes of playing time. The record base was a vinylite plastic which could withstand the wear in the very fine grooves required.

PHYSICS

313. Luis W. Alvarez. Luis W. Alvarez at the University of California produced the first successful linear accelerator (also called a Linac). By the proper phasing of the accelerating alternating voltage applied to separate cylinders, the linear accelerators have been developed to accelerate protons to almost the speed of light, developing energies on the order of 1×10^9 electron volts.

314. Patrick M.S. Blackett. The Nobel Prize this year went to Patrick Blackett of England for improvement in the Wilson cloud chamber and cosmic ray discoveries.

RADIO

315. Printed Circuits. By the end of the year, printed circuit techniques were being applied commercially. The process was pioneered by Centralab engineers and widely accepted by radio and TV manufacturers.

RADIO ASTRONOMY

316. J.G. Bolton and G.J. Stanley. J.G. Bolton and G.J. Stanley announced their discovery of discrete sources of cosmic radio noise, later known as radio stars.

STANDARDS

317. Revised Electrical Units. On January 1, the National
     Bureau of Standards and similar organizations from
     around the world adopted the system of absolute
     units derived from fundamental mechanical units.
     In this country the change amounted to only about
     0.05 percent (see 1946).

318. WWVH. The Bureau of Standards established a new
     experimental broadcast station on the island of Maui,
     Hawaii. WWVH was to cover the Pacific and Arctic
     areas which could not be reached with any consis-
     tency from station WWV in Maryland.

TELEVISION

319. Color TV. Color demonstrations were again held in
     Washington with RCA, CBS and CTI. CBS color
     was excellent, RCA was not faithful and the CTI
     system was too dim with poor registration. After
     considering the problems, the FCC favored the CBS
     system.

320 Opera. On November 29, the first full-length Metro-
     politan Opera appeared on ABC television. The
     opera was Othello and was sponsored by Texaco.

321. Earl Muntz. Earl Muntz entered the television busi-
     ness and was widely known in the United States be-
     cause of his sales promotion for television sets.
     Known as "Madman Muntz," he became a leader in
     TV-set sales but was forced into bankruptcy about
     seven years later.

322. Television Licensing Freeze. It was becoming increas-
     ingly evident that color was the coming development
     and ultimately there would be insufficient room for
     all the stations seeking licenses in the present VHF
     band. Existing sets could only receive the VHF
     channels, but most of the new stations and partic-
     ularly color would have to be in the ultrahigh fre-
     quency band where more frequencies were available
     for the wider bandwidths required by color TV.

Interference by propagation skips was also being
noted on the VHF channels. Consequently on September 29, the FCC froze all pending applications.
Authorized stations could go ahead with construction
but no others would be issued for the time being.
This freeze was to allow study of standards, frequency allocations, and co-channel interference.

TIME

323. <u>Harold Lyons</u>. The first "atomic clock" was developed
at the Bureau of Standards in 1948 and 1949 by
Harold Lyons and his associates. This was a quartz
crystal oscillator, stabilized by J=3, K=3 absorption
line in ammonia at 23,870 MHz.

<u>1949</u>

AMATEUR

324. <u>DX Records</u>. A number of VHF and UHF distance records were broken this year by the amateurs. The
new records for two-way work were these:

50 MHz--10,500 miles	CE1AH-J9AAO
144 MHz--800 miles	W3CUM-WØBIP
220 MHz--275 miles	W1CTW-VE1QY
420 MHz--262 miles	W6VIX/6-W6ZRN/6
1215 MHz--37 miles	W1OFG/1-W1MZC/1
2300 MHz--150 miles	W1IFE-W6ET/6
3300 MHz--150 miles	W6IFE/6-W6ET/6
5250 MHz--31 miles	W2LGF/2-W7FQF/2
10,000 MHz--7.65 miles	W4HPJ/3-W6IFE/3
21,000 MHz--800 ft.	W1NVL/2-W9SAD/2

325. <u>Television Records</u>. Two new records were made for
amateur television transmissions and receptions.
Using a newly designed receiver, the record was
broken on October 15, over a 17-mile path between
W2DKJ/2 and W2HID. On October 19, W2DKJ/2 was
clearly seen over a 29-mile path to W3FRE at Den-
ville, New Jersey.

COMPONENTS

326. Capacitors. A new line of capacitors was developed
by the Sprague Products Company of North Adams,
Massachusetts. These units were developed in cooperation with the American Radio Relay League and
were particularly effective for VHF bypassing. Both
metal-cased and disc capacitors were now available
with some rated at 2,500 WVDC. These units, known
as the Sprague Bypass Capacitors, were effective
for broad-band attenuation.

327. Graphecon. An oscilloscope tube called the Graphecon
was disclosed at a meeting of the IRE. The tube
can show a signal of one-billionth-second duration
for a period of one minute. A storage oscilloscope
using the tube was also exhibited at a later meeting.
The device was developed by RCA.

328. Ferro-Magnetic Material. General Ceramics and Steatite
Corp. announced the development of ferrite material
for magnetic cores. It could be molded into most
shapes and sizes with close tolerances and low losses.

329. Koch, Rutherford, and Wright. The Allen B. DuMont
Laboratories introduced its first line of metal cathode-
ray tubes for TV. These were 12- and 16-inch
models. Development has been credited to Stanley
Koch, Robert Rutherford and Gerald Wright. Sylvania soon followed in announcing their metal cathode-
ray tubes.

330. Vidicon. The development of the Vidicon by RCA opened a new field in closed-circuit television. The
small television pickup tube was highly sensitive and
permitted good contrast at normal lighting levels.
The device was quickly accepted by educational institutions and industry.

331. Tubes. A number of new tubes were released this
year which included many radio, TV and special
purpose types: DuMont--(CRT) 5XP; Eimac--592/3-
200A3, 4W1250A, 4E27A/5-125B; Electrons Inc.--1C,
3C, 6C, C1b/A, 616J; General Electric 8AP4,
GL9C24, 12AY7, 12KP4, 16GP4, GL-880, GL-5513,

GL-5610; Hytron--6BQ6GT, 25BQ6GT; National Elec-
tronics--NL614, NL653; Raytheon--1AD4, 1AE5,
6AN5, 7JP4, 8BP4, CK5744/CK619CX, CK5654,
Geiger-Mueller tubes 1B90, CK1018, CK1019; RCA--
1AC5, 1AD5, 1E8, 1T6, 3KP11, 3RP1, 4-65A,
4X150A, 4-250A/5D22, 6AB4, 6AH6, 10KP7, 12S8GT,
16AP4, 715C, 5671, 5734, 5762, 5763, 5770, 5771,
5786, 5794, 5819, 5823, 5825; Sylvania--1AC5, 1AD5,
1E8, 1T6, 1W4, 1C3, 1L6, 1B3GT, 6AG5, 6AL5,
6BG6G, 6J6, 6K6GT, 7B5, 7C5, 7F7, 7H7, 7N7, 7X6,
7Z4, 16AP4; improved tubes for TV sets--6BK6,
6GT6, 6BU6, 6K6GT, 12BK6, 12BT6, 12BU6, 26BK6,
5722; and Victoreen--1B67, 1B85, 1B87, VXR130,
V801/VX41A, 5803/VX34.

332. Noise Source. The gas-discharge noise tube was developed by Bell Labs as a means of measuring component noise and other noise of unknown origin.

333. Sylvania. By the end of the year, Sylvania had produced hermetically sealed germanium crystals in glass.

CITIZENS BAND

333a. FCC Action. In June, the Citizens Band was opened
by the FCC on the band from 460-470 MHz. There
were no technical requirements and licenses were issued upon request.

COMPUTERS

334. Edvac. The Electronic Discrete Variable Automatic
Computer (Edvac), started in 1946, was completed
and installed in the Ballistic Research Laboratories
of the Ordnance Department. It became fully operational in 1952, but had been delayed because of
personnel changes and other problems.

335. Binac. The Electronic Binary Automatic Computer
(Binac) was completed by Eckert-Mauchly Computer
Corp. and announced on August 22. It was very
high-speed compared to relay computers and each
calculation was checked by its twin computer. Oper-

ation of the two computers was in parallel with each
operation being checked. The answers had to agree
in both computers before the process could continue.

336. Wilkes, Gill and Wheeler. The Electronic Delay Storage Automatic Calculator (Edsac) went into operation
in May at Cambridge University in England. This
was the first operation of an internally stored program computer as suggested by von Neumann. The
computer was designed and constructed under the
direction of Professor Maurice Wilkes, assisted by
Stanley Gill and David Wheeler.

CONFERENCES

337. Fourth Inter-American Regional Radio Conference. On
April 25, the fourth Inter-American Regional Radio
Conference was called to order in Washington, DC.
Countries of North America, South America and Cen-
tral America participated. The meetings concluded
July 9 with the formal signing of the agreements.
The purpose of the conference was to revise the
regulations agreed to in Santiago in 1940. Ten
American republics were represented.

338- Region 2 Conference. Running concurrently with the
9. Regional Conference above was the Region 2 Conference, held in Atlantic City. Region 2 included the
French, British, and Dutch Colonies and Greenland.
Many of the same conferees were involved in both
conferences. Among the groups were representatives of the United Kingdom, Denmark, France,
Netherlands, West Indies, and Surinam.

FABRICATION TECHNIQUES

340. Danko and Abramson. Danko and Abramson of the
Army Signal Corps introduced dip soldering in 1949.
This invention greatly aided automation in electronics
which followed shortly thereafter. Details of the
process at this time were secret. When details were
released, the technique was widely used for TV and
radio chassis.

341. Motorola Laboratory. A new laboratory devoted to
     military research had been opened in Phoenix, Ar-
     izona by Motorola, Inc. of Chicago. The new lab-
     oratory was a step in decentralization of production
     units vital to national defense.

PHYSICS

342. Albert Einstein. Einstein proposed that the gravita-
     tional fields and electromagnetic fields were basically
     of the same nature, and different forms of a univer-
     sal force were proposed by Einstein. This was called
     the Unified Field Theory. To this time, there had
     been no experimental verification of this theory.

343. Hideki Yukawa. Hideki Yukawa received the Nobel
     Prize in physics for his prediction of the existence
     of mesons.

RADIO

344. Radio Free Europe. Radio Free Europe began beaming
     news towards the USSR from Munich. General Clay,
     who helped start the broadcasts, was chairman of the
     board. The station was supported by the CIA and
     by contributions from its listeners. The Voice of
     America (VOA) was authorized by the Mundt Act and
     was funded by Congress January 27, 1948.

RECORDINGS

345. Tape Recordings. Recording studios in the US began
     using tape for the first takes of musical programs
     this year.

346. 45-RPM Recordings. The 45-rpm record player was
     introduced by RCA on January 11. This was claimed
     to have been an entirely new system of reproduction
     which, with its 6-7/8-inch record, gave unsurpassed
     brilliance and clarity to the recording. The player
     included the fastest operating record-changer avail-
     able, and would provide about 40 minutes of record
     time.

TELEPHONE

347. REA Telephone Service. This year the Rural Electrification Administration (REA) began making loans to cooperatives for the installation of telephones to the farm community, much as it had done for power service earlier.

TELEVISION

348. Standards. An attempt to obtain international TV standards was made with a meeting of 11 countries in Zurich. Few standards could be agreed upon other than that of the aspect ratio, line interlacing, and that the vertical scanning should be independent of the power frequency.

349. Harry S Truman. The first telecast of a presidential inauguration was that of Harry S Truman. The telecast was carried by about 34 stations and was available to the Midwest by stratovision. An estimated 10 to 12 million viewers watched the event.

350. Kentucky Derby. The first TV telecast of the Kentucky Derby was aired in May by Station WAVE-TV in Louisville, Kentucky.

351. Tropo Scatter. The TV freeze imposed on the issuance of TV licenses in the US was caused partially by co-channel interference beyond the theoretical boundaries for the frequencies involved; this was later disclosed as tropospheric scatter.

352. Large-Scale TV. RCA demonstrated a projector TV picture 6' x 8' at the National Association of Broadcasters Convention in Atlantic City.

353. Community Antenna Television System. It was about 1949 that the Community Antenna Television Systems (CATV) began expanding around the country in areas where good TV reception was not available. The system operators installed a high-gain antenna or antennas to receive TV signals which were then amplified and distributed to the subscribers.

TIME

354. <u>WWV</u>. On November 1, the Bureau of Standards' Central Radio Propagation Laboratory expanded its service to include radio propagation notice. Standard time and audio frequencies would now be transmitted on frequencies of 2.5, 5, 10 MHz and every 5 MHz interval from 10 to 35 MHz.

355. <u>WWVH</u>. The new staiton WWVH of the Bureau of Standards was completed on the island of Maui, territory of Hawaii, and put in operaiton. The broadcasting was on an experimental basis with standard time, frequencies and musical pitch transmitted on frequencies of 5, 10 and 15 MHz.

Chapter 2

THE SIXTH DECADE: 1950-1959

As would be expected, the decade of the 1950's saw a continued development of communications, television, radio astronomy, computing and other fields. The tropospheric scatter mode of communication developed from purely experimental installations into commercial systems linking many points of the world, particularly where conventional systems were not feasible or practicable for economic or other reasons. The world was also tied closer together by a network of cables connecting many of the major countries of the world. Each cable was capable of handling many voice or data channels. This was made possible by the development of underwater amplifiers having a life of about 20 years, which were used on the deep-sea cables. Cable service in this country was also supplemented by microwave systems which, by the end of the decade, were carrying an estimated 25 percent of the long-distance telephone service in the United States. Other means of communications were also being investigated.

This period saw the beginning of satellites used for relay stations in international communications. The satellite was considered for military communication but a system was desired less subject to failure, or at least available for repair if failure did occur. An experiment of using small orbiting dipoles to reflect microwave signals was started but no experiments were made until the 1960's.

The requirement for reliable military communication in any theatre of operation was recognized. As a result, a radio system expected to provide a reliable signal anywhere on earth was put in operation by the Navy. This was a very low-frequency station with a power of over one million watts. The Army also began work on a shortwave station of about 25 million-watt power with the same objective.

Up to this time, the telephone lines other than the co-axial cables were not satisfactory for data transmission. The rise of the computer brought on more and more demand for data lines and a system known as Dataphone was developed. In this system, the data signals were converted to an analog form which could be sent over the conventional telephone network. Transmission at rates up to 800 words per minute had been obtained.

At the opening of the decade, transistors were in production but were not yet suitable for use in mass-produced electronic equipment. In general, the characteristics of identical transistor types varied so that a transistor could not be replaced by an identical type and guarantee the same circuit operation. The reason was determined to be due to impurities in the germanium. This problem was largely solved by a better method of refining the germanium. The process was known as "zone refining." A tremendous improvement in the transistor resulted and the possibilities and advantages of the transistor over the vacuum tube began to be realized. By the end of 1952 at least 25 companies had been licensed to manufacture the units.

In the zone refining process, a seed crystal of germanium would form a crystal of great purity with the impurities remaining in the melt. In the process, certain elements could be added to the melt to produce either an N- or P-type crystal. By alternating the impurities in the mix, a crystal could be formed having NPN or PNP areas which could be sliced so as to produce a number of transistors of nearly identical characteristics; these were known as "grown junction transistors."

Development continued with nearly every year bringing out new types of transistors or new or improved fabrication techniques. Types such as the field effect, surface barrier, unijunction and mesa transistors were brought out during this period. By the middle of the decade, Texas Instruments had applied zone refining to silicon. Because silicon had superior temperature characteristics, power transistors and silicon-controlled rectifiers were developed.

Initially the transistor was not a high-frequency device but suitable for audio or other low-frequency purposes. As development continued, the frequency response was increased

to where, by the end of the decade, transistors with cutoff
frequencies on the order of 500 MHz were being produced.
By the end of the decade, the transistor was showing its effect on the tube industry. Receiving tubes were in less demand. Attempts at developing a small tube to compete were
doomed to failure. Although they could be made small, they
still required filament power. As the size was reduced, the
temperature increased. The one area in which the transistor
could not compete was in transmission. Transmitting tubes
were still required and new developments and improvements
were still being made.

The electronic components during this period also under-
went a considerable period of development. The development
of the transistor itself could not produce great size reductions if the associated components were not reduced. The
lower power requirements of transistors along with the low
operating voltages and currents were a natural for size reduction. As the components reduced in size, so did the receivers, walkie-talkies and instruments in general.

While size reduction and new fabrication techniques were
being developed, other engineers were making still greater
progress in size reductions as the first integrated circuits
were developed. The first IC device on the market was the
flip-flop, a circuit widely used in digital computers. The
use of discrete components in electronic circuitry appeared
to be rapidly drawing to an end.

With the development of television, interest in both AM
and FM broadcasting declined; however, predictions of television putting the radio broadcasters out of business did not
develop. In 1945 the Federal Communications Commission had
moved FM broadcasting to the 88 to 108 MHz band. As a
result, the FM stations could no longer be received on the
old receivers, further slowing FM progress. Broadcasters
had continued to apply for FM licenses in the event that FM
should develop into good business or replace AM stations.
This trend continued until about 1948 when it peaked. The
advantages of FM were not promoted and by 1950 most FM
stations were carrying the same programs as the associated
AM stations.

After the initial excitement of television had declined, the
interest in high-fidelity music revived an interest in FM

broadcasting. The development of stereo transmission and
the ability to carry both stereo channels on one carrier frequency sparked the FM broadcasting revival.

About this time color television was coming to some of
the larger cities. The electro-mechanical color receivers of
the Columbia Broadcasting System (CBS) could not receive
black-and-white pictures and vice versa. RCA had developed a system of color which could be seen either in color
or in black and white. After a stormy time in which the
Federal Communications Commission adopted the CBS system
as the standard for set manufacturers, the FCC was forced
to issue a "stop" order to manufacturers to allow time for
further consideration of the problem. In the meantime, RCA
continued improving their system, and with the newly developed color tubes, excellent reproduction of color and
black-and-white pictures was possible. As a result, the
FCC was forced to reverse its initial decision and adopt the
RCA method of color as standard for the industry.

It was during this decade that the total income of television stations exceeded that of the broadcasting stations.

In 1957, the Soviet Union put the first man-made satellite in orbit. The idea of using an orbiting satellite as a
relay station had previously been proposed. By the end of
the following year the US Army had put in orbit a satellite
capable of relaying either telephone or teletype signals.
This first satellite did not operate long, but long enough to
demonstrate the feasibility of satellite communication.

Another method of communication was demonstrated, per-
haps more as a stunt than for practical purpose of communication, by bouncing a signal off the moon to be picked up
at some remote point on earth.

Some amateur stations also succeeded in picking up their
own signals after they were reflected from the moon. This
was a real accomplishment because the legal power limit for
the amateur was 1,000 watts. Another ham was successful
with only 700 watts. While these tests were ongoing, other
amateurs, realizing the value of the service amateurs could
render in times of emergency, organized the Radio Amateur
Civil Emergency Service (RACES), an organization dedicated
to the service of aiding those persons caught in some emer-

gency situation. A further achievement of the amateurs and a new "first" was the accomplishment of transatlantic television.

The era of the 1950's was a period of improving the all-electronic computers. By the middle of the decade there were many computers on the market and in use by a number of businesses and organizations. During this time the computers in operation included Seac, SSEC, Tridac, Univac, Erma, Norc, Ramac and others by RCA, IBM and Philco. Programming of the computers had always been a rather complicated and time-consuming process. This problem was greatly reduced by the development of ALGOL and COBOL, simplified computer languages easily understood by the computer and the programmer. The computer operations were being simplified to where they could be operated by nearly anyone after a relatively short training period. The transistor was being applied to the computer circuits resulting in the first all-transistorized computer.

The decade saw the increased popularity of the microwave ovens to the public. Automatic control was also being applied to instruments; first with the automatic tube characteristics plotter and later with an automatic balancing bridge circuit; however, the most outstanding developments in the decade were the development of the maser and the proposal of an optical maser, both outstanding achievements having many applications in the coming years of space exploration and satellite communication now on the horizon.

In the field of radio astronomy, it was discovered that the planets, the sun, moon and comets were emitters. Radio frequency bursts from Jupiter and Venus were detected. Several new observatories were started. The idea that there may be intelligent beings in the galaxy or beyond trying to communicate with us took hold and at least one listening post was started with the aim of determining if any such signals from an intelligent source could be detected.

Atomic power became a reality in this period with Russia putting one of the first plants in operation in 1954. Before the decade was out atomic power systems had been put in operation in England and Japan. Sweden was supplying power to the island of Gotland by a 100 kv DC submarine cable which was later converted to AC by a mercury arc converter.

1950

AMATEUR

356. 144-MHz Record. An official distance record for 144 MHz of 1,200 miles was set in June by stations W5VY and W8WXV.

AUTOMATION

357. Project Tinkertoy. The Bureau of Standards was asked by the Navy Bureau of Aeronautics to develop the automation of circuit assemblies for electronic devices. The resulting Project Tinkertoy led to the first modular packaging.

BROADCASTING

358. Broadcasting. With the development of television, the interest in frequency modulation declined for awhile, and many stations went off the air. The country was turning to TV.

COMMUNICATION

359. Telex. Transatlantic teleprinter service (Telex) was started between New York and Holland. Arrangements were made between the Netherlands Telecommunications Administration and Radio Corporation of America.

COMPONENTS

360. Gordon Teal. Transistors were being produced by this time; however, impurities in the germanium made it very difficult to produce units of identical characteristics. The zone refining technique for purification of the germanium, developed at Bell Laboratories by Gordon Teal, made possible the drawing of germanium with controlled impurities to

provide the desired characteristics for transistor or
diode use. By changing the nature of the impurity
in the germanium, either N- or P-type germanium
could be produced. The system made the grown
junction transistor manufacture become practicable.

361. R.N. Hall and John Saby. Doctors R.N. Hall and
John Saby of the General Electric Laboratories in
Schenectady, built the first alloyed PN junction using indium-doped germanium. This led to the alloy
junction transistor.

362. Tubes. Among the new or updated versions of vacuum
tubes advertised this year were: Amperex--492-R/
5758, AX9900/5866, AX9901/5867, AX9902/5868,
AX9903, AGR9950/5869, AX9951/5870 (the highest
power air-cooled tube built by Amperex and one
of the highest in the world at this time is the
AX9906/6078; it is rated at 108 kw output); Eimac--
2C39A; Electronic Tube Corp.--55J6 (a 5-gun CRT);
Federal Telephone and Radio Corp.--F-5918, F-5512;
General Electric Co.--GL4D21/4-125A, GL4D250A/
5D22, 6AS5, 6BC5, 6BQ6, 25BQ6, GL1614, 58844,
GL6019 (Ceramic), GL-5855, 6W6GT; Heintz & Kauf-
man--450TL/HK854L, HK257B; Hytron--6U4GT, 6AU5,
12GH7; Ken Rad--6AL7GT, 6E5, 6U5, 12AT7; National
Electronics--NL635; National Union--5851; Raytheon--
5829; RCA--6AX5, 5675, 5826, 5831, 5946; Sylvania
--1AF4, 1AF5, 1U6, 1V2, 1X2, 3E5, 6AB4, 6AB5,
6AD4, 6BQ6GT, 6L6WGA, 6SL7W, 6SN7W, 6X5WGT,
12AY7, 28D7W, 5645, 5646; Tungsol--6AX4GT; and
Victoreen--5841.

363. 500-kw Tubes. A tube rated at 500 kw continuous
operation, the world's most powerful, was announced
by RCA in February.

364. RCA Color Tube. In March, RCA demonstrated a color
tube having 351,000 dots of three different color
phosphors which permitted a color set compatible with
black-and-white standards to be built. On April 6,
RCA notified the FCC that it had developed a compatible all-electronic color TV operating in a 6 MHz
channel. RCA later demonstrated compatible color
TV to the FCC.

365. <u>Paul Weimer, Stanley Forgue, and Robert Goodrich</u>.
     Announcement of the development of the Vidicon TV
     camera tube was made in May be Weimer, Forgue
     and Goodrich of RCA. The Vidicon was the first
     camera tube to use photoconductivity rather than
     photoemission in its operation.

366. <u>Relays</u>. Reed relays, giving greater switching speeds
     than the conventional telephone relays, were de-
     veloped at the Bell Telephone Laboratories.

COMPUTERS

367. <u>Seac</u>. The Standards Electronic Automatic Computer
     (Seac) was put in operation this year at the Bureau
     of Standards, Washington, DC. It was a one mega-
     pulse/second machine with a word length of 44 bits.
     This was the first US machine to use cathode-ray
     tubes as storage devices.

INDUSTRY

368. <u>Eckert-Mauchly Computer Corporation</u>. Loans due the
     Totalisator Company were due in January and again
     the Eckert-Mauchly Corporation was having financial
     difficulties. Remington Rand, Inc. agreed to pay
     the loans and take over the corporation. This was
     the salvation of Eckert and Mauchly, and it marked
     Remington Rand's entry into the digital computer
     field. The Eckert-Mauchly Computer Corporation
     continued development of the Univac as a subsidiary
     of Remington Rand.

INSTRUMENTATION

369. <u>Tube Characteristics Plotter</u>. The National Bureau of
     Standards developed an instrument which gave an
     instantaneous visual display of vacuum tube charac-
     teristics. The plate-current vs. plate-voltage curves
     were plotted on the screen of a cathode-ray tube.
     The current and voltage scale was all displayed with
     the curves. Overall accuracy was within ± 5 percent.

LITIGATION

370. Compatible TV. RCA filed suit against the FCC decision on noncompatible color TV with the US District Court of Chicago. The Court upheld the ruling of the FCC. The appeal was carried to the US Supreme Court which ruled the following year to uphold the decision of the District Court.

MAGNETIC BEARINGS

371. Jesse W. Beams. Professor Jesse W. Beams, at the University of Virginia, developed a 16,600 rpm centrifuge motor using photoelectric positioning to maintain axial stability at the high speed.

MEDICAL

372. Russell H. Morgan. On December 5, Dr. Russell H. Morgan demonstrated his TV-screen intensifier and performed the first intercity medical diagnosis by television. This screen increased the contrasts and density of X-ray film to permit TV-transmission medical use.

NOBEL PRIZE

373. Cecil Frank Powell. C.F. Powell of England won the Nobel Prize for developing a photographic method for studying nuclear processes.

PATENTS

374. John Bardeen and Walter Brattain. The patent for the point contact transistor was issued to John Bardeen and Walter Brattain in October, and assigned to the Bell Telephone Laboratories. (See also Electrical and Electronic Technologies; 1900-1940, entry no. 886.)

PHYSICS

375. X-ray Microscope. The first X-ray microscope was
made public, giving much finer resolution and consequently greater detail than the conventional microscope.

376. T.E. Allibone. A program of holographic electron microscope development was started in the research
laboratory of the Associated Industries in Alder-
maston, England under the direction of Dr. T.E.
Allibone. Work was eventually discontinued, primarily because no adequate source of coherent light was
available at this time.

RADIO

377. Round-the-World Signals. The National Bureau of
Standards detected very low frequency radio signals
which were sent from station NSS in Annapolis,
Maryland and received after having apparently travelled around the world. NSS used a power of 350
kw on 18 kHz. The signal delay was 0.125 seconds.

378. Clock Radio. By this year, the clock radio had given
a shot-in-the-arm to the dying radio industry. Over
seven million broadcast sets were manufactured this
year but not all were clock radios.

379. Pocket Radio. A superheterodyne receiver complete
with batteries and speaker, yet small enough to fit
in a pocket, was developed at the RCA Laboratories
and announced on March 9.

380. Vacuum Tubes. At the beginning of the year, approximately 40 manufacturers were making vacuum tubes.

RADIO AIDS TO NAVIGATION

381. Loran Tower. An experimental Loran tower was completed in Forestport, NY, built by the Air Force
Electronic Center at Griffiss Air Force Base. The
tower was over 1,200 feet high, making it the second

tallest man-made structure (the first being the Empire State Building). It was supported by 18 steel guy cables.

TELEPHONE

382. Bell System. The microwave relay stations were put in service this year by Bell. The TD-2 microwave system was put in service between Chicago and New York. By 1972, the systems could carry 1,500 telephone channels on each of the 12 broadband channels. The type-N carrier system was also introduced this year. The line capability was 12 circuits, but new methods were used to reduce noise and cross talk. It was intended primarily for short-distance operation.

383. Telephone Repeaters. The first telephone repeaters for deep-sea underwater lines were installed in cables between Key West and Havana, Cuba this year. Others were in use but in more shallow water.

TELEVISION

384. International Radio Consultative Committee. The International Radio Consultative Committee met in London to consider television standards. Twenty-two nations attended but it was concluded that it was too early for color standards to be adopted at this time.

385. Cable Systems. By the end of the year, there were 70 known TV cable systems in operation. These systems began because of the few TV stations and the many areas where TV could not be received.

386. President Truman. Truman's presidential message to Congress this year was the first to be televised.

387. Peter C. Goldmark. In August, Peter Goldmark of the Columbia Broadcasting System (CBS) demonstrated his field sequential color television system. The system used three revolving color filters. A color field was taken through each color filter sequentially.

These three color fields were separated by a rotating filter at the receiver. The color picture frequency was 20 per second with 120 fields per second. The system used 375 lines and was not compatible with black-and-white sets.

388. TV Comparison Tests. In side-by-side tests before the FCC the color systems of RCA and CBS were compared. At this time the system of the CBS appeared to give somewhat better color. It did have two major weaknesses compared to the RCA system. First, it had a mechanical color filter which, for large-screen television pictures, would be roughly three times or more the screen size in diameter. Second, it was not compatible with the existing black-and-white sets, so color pictures could not be seen in black and white. New sets would have a switch to receive the desired picture. In spite of these problems, in October the FCC adopted the CBS standards for the industry, effective November 20.

389. Coaxial Transmission. RCA demonstrated color television sent over coaxial cable between New York and Washington.

TIME

390. Precision Time. At a meeting of the International Scientific Radio Union in Zurich, it was reported that quartz clocks had achieved the stability over long periods to reveal variations in the rate of rotation of the earth.

391. Time Broadcasts. The services of WWV were altered and a new series inaugurated on January 1. New services would include standard radio frequencies of 2.5 MHz and at every 5-MHz interval from 5 to 35 MHz inclusive. Time announcements would be made at five-minute intervals by voice and CW, with standard time intervals of one second. Standard frequencies of 440 and 600 Hz and propagation disturbance warnings were also transmitted. WWVH, still in experimental status, was broadcast on 5, 10 and 15 MHz with services basically the same as WWV.

1951

AMATEUR

392. <u>144-MHz Record</u>. On June 10, an unusual opening in the 144-MHz band permitted W6WSQ in Pasadena and several other California stations to work W5QNL, Texarkana, Texas. This was 1,200 to 1,400 miles and a new DX record for two meters.

AUTOMATION

393. <u>Robert Henry</u>. Project Tinkertoy, led by Robert Henry of the Bureau of Standards, developed a new form of automatic assembly and wiring of components into a complete module. By 1957 the packaging ideas had shifted to miniaturization.

BROADCASTING

394. <u>Remote-Control FM Stations</u>. In 1951 the Federal Communication Commission authorized the use of remote control for low-power educational FM stations.

COMMUNICATION

395. <u>Tropospheric Scatter</u>. Tropospheric scatter tests were conducted in January by Collins Radio Co. using a 20 kw 49.8 MHz transmitter. The signals were easily heard on the east coast in Sterling, Virginia--a range of over 700 miles. Rhombic antennas were used. The signals averaged 16 db over 1 μv (microvolt) about 50 percent of the time.

396. <u>Moon Bounce Communication</u>. On November 8 the National Bureau of Standards sent a message via moon bounce from Collins Radio Co., Cedar Rapids, Iowa, to Sterling, Virginia, on 418 MHz. Power was 20 kw.

COMPONENTS

397. **Shockley, Sparks and Teal**. The first description of the possibilities of the junction transistor was in an article by William Shockley, Morgan Sparks, and Gordon K. Teal in the **Physical Review**, vol. 83, 1951.

398. **W. Shockley**. The first practical field-effect transistor (FET) was invented by William Shockley of the Bell Telephone Laboratories.

399. **Tubes**. Among the new tubes brought out this year were: Amperex--AX9903/5894; Eimac--4W 20, 000A, 4-100; General Electric--Z2061, GL-6039; National Electronics--NL643; RCA--17CP4, metal rectangular CRT; and United--576, 577, 578, 371B, 3B24W, 3B29.

400. **Plasmatron**. The development of the plasmatron gas discharge tube was announced at the IRE National Convention on March 21. The tube provided a high-speed means of power and circuit control.

401. **Digital Readout Tubes**. The first digital readout tubes were brought out this year by Haydu (the Nixie Tube) and National Union (Inditron). These tubes permitted digital instruments such as frequency and pulse counter, voltmeters, etc. to be developed.

COMPUTERS

402. **Whirlwind I Computer**. The Whirlwind I computer, built at MIT, was installed in March at the Digital Computer Laboratory. Initially storage was made in cathode-ray tubes, with each tube storing 1,024 binary digits. These tubes were slow and development was continuing on a magnetic memory. In 1952, the new magnetic memory was showing access time of one microsecond or less. The computer used about 500 tubes and 11,000 diodes.

403. **Univac I**. The Universal Automatic Computer (Univac) was built for the Census Bureau and put in operation

in April. It was built by J. Presper Eckert and
J.W. Mauchly, now with Remington Rand. This is
considered one of the best of the first-generation
computers. It used 5,000 vacuum tubes. Univac I
was the only mercury delay storage computer to become commercially available by this time.

404. Eniac Computer. The Eniac was moved to the Ballistic
Research Laboratories at Aberdeen, Maryland, where
it operated until October 2, 1955 when it was retired
from active service.

INDUSTRY

405. Mutual Broadcasting System. Control of the Mutual
Broadcasting System was taken over this year by
the General Tire and Rubber Company, which purchased the controlling interest of MBS.

406. CBS-Hytron. The Hytron Radio and Electronic Corporation was bought by the Columbia Broadcasting
System and entered into the manufacturing business
as CBS-Hytron. Hytron had long been a well-known
tube manufacturer. Later in the year the CBS system was reorganized into CBS-Columbia, CBS-Hytron,
CBS Radio, Columbia Records, CBS Television, and
CBS Laboratories.

407. Texas Instruments, Inc. The Geophysical Services,
Inc., specialists in geophysical exploration, were
entering the electronics field and reorganized under
the name of Texas Instruments, Inc.

LITIGATION

408. Color TV. Following the adoption of the CBS standards for color TV the manufacturers delayed tooling
up for the new standards, realizing no mechanical
and incompatible system could endure. RCA filed
suit against adoption of the new FCC rules. On
May 28, the case came up before the Supreme Court
which upheld the FCC decision. About that time,
manufacturers were asked to suspend the manufacture
of color sets for the duration of the Korean War.

MASERS

409. V.A. Fabrikant. The recognition of the possibility of the amplification of electromagnetic radiation by stimulation was made by V.A. Fabrikant in Russia. He applied for a Soviet patent this year but so far as is known, he did not produce an operating maser.

MEDICAL

410. M.G. Ananev. Dr. M.G. Ananev in Russia tried a combination of AC and DC voltage and produced safe narcosis of animals. The return to consciousness brought none of the aftereffects common with drug or gas anesthesia.

ORGANIZATIONS

411. National Association of Radio and Television Broadcasters. The National Association of Broadcasters merged with the Television Broadcasters Association. The merger became the National Association of Radio and Television Broadcasters.

PATENTS

412. William Shockley. The patent on the grown junction transistor was issued to William Shockley this year.

PHYSICS

413. John Douglas Cockroft and Ernest T.S. Walton. Work on artificially produced changes in atomic nuclei gave the Nobel Prize to J.D. Cockroft of England and E.T.S. Walton of Ireland this year.

414. Ion Implantation. The technique of ion bombardment in modifying properties of certain materials was discovered during the late 1940's or early 1950's. The technique of ion implantation was disclosed at the Bell Laboratories this year. Ion implantation became

a widely used technique later in developing solid-state devices.

POWER

415. Atomic Power. On December 1, the first-known generation of electric power from atomic energy was produced by the Experimental Breeder Reactor- #1 (EBR-1), which operated trouble-free for four years. EBR-1 was developed by scientists of the Argonne National Laboratory of the Atomic Energy Commission in Idaho. EBR-1 was set up because of uncertainty in the future supply of uranium ore. EBR-1 was built to provide data on breeder possibility in power units. EBR-1 was shut down December 31, 1963 after having demonstrated the feasibility of breeding and generating useful power.

RECORDING

416. Video Recorder. The first demonstration of a video tape recorder in this country took place on November 11. It was made by Bing Crosby Enterprises; the display was in black and white.

SONAR

417. Wayne M. Ross. The fish-locater, operating on the principle of sonar was developed by Wayne M. Ross. An ultrasonic beam could be directed in any direction under the surface and echoes were displayed on an electronic screen or as sound. With practice, the characteristic echoes from different objects could be recognized.

TELEPHONE

418. Direct Dialing. Direct long-distance dialing was introduced by the Bell System this year. Both coasts were now linked by a microwave system for the telephone.

TELEVISION

419. TV Production Halt. On October 19, by direction of
the Director of Defense Mobilization, a halt was
called to the production of color television sets for
the duration of the Korean War.

420. Color Casting. The Columbia Broadcasting System
telecast the first commercial colorcast in TV history.
It was an hour-long program aired June 25. The
picture could not be received on an unmodified
black-and-white set. It was estimated that there
were fewer than 50 color sets in the United States
of the type required for seeing the telecast. It
was seen on some modified black-and-white sets,
however.

421. NTSC Standards. In November, the National Television System Committee (NTSC) approved specifications standards for compatible color TV.

422. Coast-to-Coast TV. The microwave radio relay system
built by AT&T made possible the first coast-to-coast
television operation. The occasion was President
Truman's opening of the Japanese Peace Treaty Conference in San Francisco.

WHISTLERS

423. L.R.O. Storey. L.R.O. Storey began the study of
whistlers. He discovered that the whistlers apparently followed along the lines of force of the earth's
magnetic field.

1952

AMATEUR

424. Edison Award. In October, the Edison Radio Amateur
Award was announced by the General Electric Company. The award was set up to gain greater recog-

nition for those radio amateurs performing meritorious public service in disasters, civil defense or other emergencies.

425. 21-MHz Band. The problem of frequency allocation below 27.5 MHz had been growing since original assignments were made in 1927. A new start had been made in 1947 to analyze the requirements of the various services needing the higher frequencies and to allocate frequencies and bands on a carefully engineered basis. After assignments were made, they would be approved by an international conference. The international treaty, written in Atlantic City, specified the band of 21,000 to 21,450 kHz for the amateur who would also lose 50 KC of the 20-meter band. The agreement came only after many delays and conflicts. By March, the conflicts had been resolved. A Geneva Conference in 1951 ratified the plan but no specific dates were set as to when it would be effective. The official opening date for the 21 MHz band was set for May 1, for cw use only.

426. RACES. The Radio Amateur Civil Emergency Service (RACES) was inaugurated to provide almost instant warning in case of emergency from enemy action or other disasters.

COMPONENTS

427. Capacitors. The solid-state capacitors using tantalum oxide dielectric were developed by Bell Labs. Similar to electrolytic capacitors in operation, but without the liquid, they provided more capacity per unit volume and were particularly applicable to the development of miniature and subminiature circuits.

428. Transistors. By September, approximately 25 domestic and nine foreign companies had been licensed by the Western Electric Company to manufacture transistors.

429. RCA. Point contact transistors capable of oscillating at frequencies up to 200 MHz were announced on June 20 by RCA. Some of the transistors could handle 500 mw of power.

430. Tubes. Television was "growing up." A new 21" TV
     picture tube was brought out by RCA. Other new
     tubes included: Eimac--4X150D; General Electric--
     6AJ4, 6AM4, 6BK5, 6BX7GT, 6U8; and Raytheon--
     6146.

431. Marlin E. Bourns. The first of a long line of minia-
     ture potentiometers with leadscrew adjustments,
     known as the Trimpot, were introduced by Marlin
     E. Bourns at the Instrument Society of America Show
     in Cleveland, Ohio. The Bourns organization had
     become a leader in resistive components with a world-
     wide reputation and sales organization.

     COMPUTERS

432. Princeton Computer. The Princeton Institute for Ad-
     vanced Study Computer was completed this year,
     and was the first computer to use parallel arithmetic.
     It had what was considered an unusually large mem-
     ory at this time--1,024 words. The project director
     was John von Neumann.

433. Selective Sequence Electronic Calculator (SSEC). The
     Selective Sequence Electronic Calculator, completed
     in 1948, was reported to have been scrapped this
     year. It had performed over 1,500 man-years of
     calculations in one 150-hour period.

434. Bizmac. The Bizmac business computer was introduced
     by RCA. It was one of the first variable word-
     length computers. The computer used both storage
     tubes and magnetic tape storage. One of its first
     applications was in inventory control. In this use
     as many as 200 tape stations with tapes mounted
     were made available for immediate processing.

435. Grace Hopper. A major step in computer programming
     was made by Lieutenant Grace Hopper of the Navy
     for the Univac. This device, known as the A-O
     Compiler, took the computer program written in al-
     gebraic notation and converted it to machine language
     understood by Univac.

FABRICATION TECHNIQUES

436. Printed Circuit Boards. The first copper-clad epoxy-
glass laminate, used later for printed circuits, was
developed this year by PCK Technology. The method of plating through holes for connecting conductors
on both sides of the board was developed the next
year.

INDUSTRY

437. Andy Kay. Industry's "Digital Revolution" started
this year with the first digital voltmeter displayed
by Andy Kay. The model-419 sold for $4,000. The
accuracy was on the order of .01 percent. Non-
Linear Systems was organized to manufacture and
market digital meters.

438. Texas Instruments, Inc. Texas Instruments entered
the semiconductor field and obtained a license to
manufacture transistors from the Bell Telephone
Laboratories. They brought out the first silicon
transistor about two years later.

INSTRUMENTATION

439. Frequency Counter. Hewlett-Packard brought out its
first frequency counters with a capability of counting at rates up to 10 MHz with a resolution of about
.01 Hz.

INTEGRATED CIRCUITS

440. George W.A. Dummer. At the annual Electronic Components Symposium in Washington, DC in May,
George W.A. Dummer of the Royal Radar Establishment of Great Britain and one of the developers of
the Plan Position Indicator (PPI) for radar, read a
paper in which he presented the first idea of integrated circuits. At the closing of his paper, he
made the prediction:

With the advent of the transistor and the work in
semiconductors generally, it seems now possible
to envisage electronic equipment in a solid block
with no connecting wires. The block may consist
of layers of insulating, conducting, rectifying and
amplifying materials, the electrical functions being
connected directly by cutting out areas of the
various layers.

MEDICAL

441. Hearing Aids. The first hearing aids to utilize transistors for a reduction in size were developed this
year by Sonotone. They required a tube for the
input and driver stage, however.

PHYSICS

442. E. Courant, M.S. Livingston and H. Snyder. Scientists at the Brookhaven National Laboratory proposed
magnetic beam focusing. Protons were focused to a
narrow beam in a vacuum chamber to produce the
focusing electron synchrotron. This accelerator
permitted cheaper and smaller accelerators to be
built.

443. Oak Ridge National Laboratory. A 63-inch cyclotron
was put in service in May at the Oak Ridge National
Laboratory. This cyclotron was designed to accelerate nitrogen ions to approximately 27 MeV.

444. Edward Mills Purcell and Felix Bloch. Edward Purcell
and Felix Bloch, both of the United States, received
the Nobel Prize for measuring the magnetic fields in
atomic nuclei.

RADAR

445. SS Normandie. The French made the first commercial
application of radar with an obstacle detector, using
a form of radar, installed on the SS Normandie.

TELEPHONE

446. Automatic Answering Sets. The first coded automatic
    answering sets were developed in 1952 by the Bell
    Telephone Laboratories. The type-1 answering sets
    recorded on magnetic drums and included a variable
    announcement cycle.

TELEVISION

447. TV Ban Lifted. The ban on TV station construction,
    in effect since 1948, was lifted on April 14. Eighty-
    two television channels were now provided for.
    Twelve VHF channels were currently in use. Seven-
    ty more channels in the UHF band between 470 and
    890 MHz were now available with 390 stations author-
    ized within the next year. These were the uncon-
    tested applications.

448. Gerald Herzog. The first all-transistor TV was de-
    veloped this year by Gerald Herzog, a research en-
    gineer at RCA.

1953

AIR CONDITIONING

449. Heat Pumps. The air conditioning units having the
    capability of reverse-cycle heating became commer-
    cially available about this time. These so-called
    heat pumps became especially popular for new con-
    struction projects.

AIRPORT LIGHTING

450. Airport Lighting. An emergency system had been de-
    veloped which permitted a pilot in trouble to turn on
    the lights of nearby small airports from the air.
    The first field to be equipped for the service was
    the airport at Kingman, Arizona.

AMATEUR

451. <u>Radio Astronomy</u>. In the spring, W4AD bounced sig-
     nals off the moon which were received by W3GKP
     and W3LZD. This effort was started in 1950. The
     operation was on 144 MHz and the transmitter 1 kw.

452. <u>Transistor to Transistor Communication</u>. The com-
     munication between W2JEP and W2YTH is believed to
     have been the first between transistorized transmit-
     ters. The contact was made on February 19 at a
     range of about one-half mile. The transmitters op-
     erated on the 40-meter band with about 60 milliwatts
     input.

453. <u>21-MHz Band Privileges Expanded</u>. Effective at 3 AM
     EST March 28, the FCC revised its regulations to
     permit A-3 (phone) and narrow-band frequency
     modulation in the 21 MHz band, 21,250 to 21,450 kc
     (kilocycles). This was for operators other than
     those with novice or technician licenses.

BROADCASTING

454. <u>American Broadcasting Company (ABC)</u>. With the pur-
     chase of the Blue network of the National Broadcast-
     ing Company, the American Broadcasting Company
     became weak because of its lack of advertising and
     program material and began to slip financially. With
     the approval of the Federal Communications Commis-
     sion, ABC merged with, and came under control of
     United Paramount Theatres. This action allowed ABC
     to recover.

CIRCUITS

455. <u>Filters</u>. The Collins Radio Corporation introduced
     their new mechanical filter. The filter was extreme-
     ly sharp with a band-pass of only a few hundred
     cycles. This made the reception of stations in the
     crowded bands possible to an extent not previously
     achieved.

COMMUNICATION

456. <u>Tropospheric Scatter</u>. Construction of the first UHF
over-the-horizon communication system was started
between Baffin Island and Newfoundland. The system provided 35 voice channels operating between
800 and 900 MHz. Stations were located at intervals
of 100 to 200 miles, depending on the terrain.

COMPONENTS

457. <u>B.Y. Mills and A.G. Little</u>. The Mills Cross antenna,
later widely used by Radio Astronomy Laboratories,
was first described by B.Y. Mills and A.G. Little
in the <u>Australian Journal of Physics</u>, Vol. 6.

458. <u>Tantalum Capacitors</u>. The microminiature tantalum
capacitors were introduced by the General Electric
Co.

459. <u>Surface Barrier Transistor</u>. High-frequency transistors
came on the market with the Philco Surface Barrier
Transistor. Frequencies in the megahertz region
could be amplified. The transistor was made using
the jet etching techniques. This high frequency
was accomplished by etching the base material to a
very thin surface which was plated on both sides
with indium metal to form the emitter and collector.

460. <u>I.A. Lesk</u>. The Unijunction transistor, developed by
I.A. Lesk at the General Electric Advanced Semiconductor Laboratories, came on the market. This
was particularly suitable in switching circuits and
was a bar of N-type germanium with a P-type area
diffused in the center of the bar serving as the
emitter control element; an ohmic contact on each
end of the base material served as the switching
contacts.

461. <u>Tubes</u>. Among the tubes announced this year were:
Amperex--6252; Eimac--4W300B; General Electric--
GL5899, GL6202, GL6203, GL6021, GL6134; Los Gatos
--705A; RCA--5690, 5719, 5840; and Thermosen, Inc.
(Kalotron)--5DC-5, 5DC-5M.

462. Jay Forrester and Bill Papian. Magnetic bead mem-
      ories were developed under Project Whirlwind at
      MIT by Bill Papian under the direction of the in-
      ventor Jay Forrester. This device nearly tripled
      the speed of access to the memory over that of the
      storage tube. In August, the first bank of core
      storage was wired into the Whirlwind computer.
      Operation was improved to the point where develop-
      ment work on storage tubes was halted.

      CONELRAD

463. Control of Electromagnetic Radiation (Conelrad). Un-
      der presidential authority, and with the cooperation
      of the FCC a plan was put into effect to confuse
      aircraft or guided missiles attempting to home in on
      a particular broadcast station. The first Conelrad
      test was held September 16 with stations switching
      programs from one transmitter to another in a simu-
      lated attack. Over 1,700 stations took part in the
      tests.

      CONFERENCES

464. International Conference on Atomic Power. The First
      International Conference on Atomic Power was held
      August 11-13 in Oslo, Norway. Scientists from 19
      countries attended.

      ELEVATORS

465. Otis Elevators. Automatic elevator service was demon-
      strated by the Otis Elevator Company. This de-
      velopment soon eliminated the necessity for the ele-
      vator operator.

      FABRICATION TECHNIQUES

466. Wire Wrapping. The technique of wire wrapping for
      the construction of electronic circuits was introduced
      by R.F. Mallina of Western Electric Company about

this time. The wire was wrapped around a square pin terminal so tightly that soldering was unnecessary.

467. Project Tinkertoy. A semiautomatic method of electronic fabrication was developed at the National Bureau of Standards. The system permitted the time of equipment manufacture to be considerably shortened. The method was developed under the name of Project Tinkertoy. The project was sponsored by the Navy.

INDUSTRY

468. Gordon Teal. Texas Instruments, Inc. set up a semiconductor laboratory headed by Gordon Teal. They were producing junction and point contact transistors by the end of the year.

ORGANIZATIONS

469. Antique Wireless Association. A group of amateur historians, interested in the history of both electrical and electronic communications, organized the Antique Wireless Association in Holcomb, New York. The organization grew into an internationally recognized organization with members in the US and in other parts of the world.

OSCILLOSCOPES

470. Tektronix. The first plug-in oscilloscope, pioneered by Tektronix, came on the market. Model 535 was available with three types of plug-in preamplifiers. One plug-in permitted dual channel operation.

POWER

471. Atomic Reactor. The first privately operated atomic reactor was opened in May by North Carolina State University and the University of North Carolina. It was set up without government financial support.

472. Nuclear Power. Westinghouse was assigned the project for the design and development of the first full-scale nuclear power plant in the country. An output of 60 million watts was anticipated. The installation was to be located at Shippingport, Pennsylvania.

RADIO

473. Radio Jim Creek. The Navy's long-wave 1,200,000-watt output transmitter in the Cascade Mountains of Washington was officially put on the air on November 18. The station, started in 1946, would be able to reach naval units anywhere on earth without propagation problems characteristic of the higher frequencies. The first official message was sent by David Sarnoff.

RECORDING

474. Video Recording. On December 1, RCA demonstrated its video tape recording system at the David Sarnoff Research Center. This recorder could provide black-and-white or color pictures. The tape moved at 30 ft/sec. and provided a 3-MHz bandwidth. Two tracks were used for black-and-white pictures; one for the video signal and synchronizing pulses and the other for sound. This was on 1/4-inch tape. Color TV was recorded on a half-inch five-track tape. Each color required a separate track, one track for synchronizing signals, and one track for sound.

TELEPHONE

475. Automatic Answering Service. The Bell Telephone Company publicly announced a device which, when the telephone rings, would answer with a prerecorded message and then stand by to record an answer. After 25 seconds, the machine would reset.

476. 50-Millionth Telephone. The 50-millionth telephone was installed this year by the Bell System.

TELEVISION

477. Compatible TV. On June 25, RCA and NBC petitioned
the FCC to adopt the compatible signal specifications
used by RCA as standard for color television. The
specifications had previously been approved by the
National Television Systems Committee.
On October 15, RCA and other members of the
TV industry demonstrated compatible color TV to the
FCC and showed that it was ready for the public.
On December 17, the FCC withdrew its 1950 approval
of the CBS color system in favor of the compatible
all-electronic system developed by RCA. TV-set
manufacturers were given details of the basic system.

478. Educational TV. The first educational TV station to
go on the air was reported to have been KUHT in
Houston, Texas.

479. Station Ownership. By a ruling of the FCC, the number of stations that any person or group of persons
could own was set at five for TV and seven for AM
or FM. The next year the television limit was
raised to seven for TV.

1954

AMATEUR

480. 10,000-MHz Record. A new amateur record in the
United States for 10,000-MHz communication was
made March 13. Communication was over a 22.8
mile path between W7J1P/7 on Mt. Scott near Port-
land, Oregon and W70KC/7 at Rocky Point on the
Columbia River. Converted and home-built equipment was used. The British record at this time was
27 miles. Not to be out-done, on April 10 they
worked over a 47-mile path for a new world's record
for the frequency.

COMPONENTS

481. **Variable Vacuum Condensers**. The VAC-4-40 vacuum variable condenser, manufactured by the Jennings Radio Mfg. Corp. of San Jose, California, became available early in the year. The high price limited its application for amateur use.

482. **Chapin, Fuller and Pearson**. Silicon Solar Cells were invented by D.M. Chapin, C.S. Fuller and G.L. Pearson of Bell Labs. The cells convert sunlight directly into electricity and are widely used for powering electronic devices not readily accessible, such as satellites, remote beacons, sea buoys, etc.

483. **Klystron**. A new Klystron tube was brought out by the Sperry Gryoscope Co. This tube could provide an output of about four-million watts and was eight feet long.

484. **Resistors**. Sprague introduced its Blue Jacket line of resistors. These were three-watt wire-wound units about the size of a one-watt carbon resistor.

485. **Zenner Diodes**. The first Zenner diodes were put on the market by National Semiconductor, a division of National Fabricated Products Co.

486. **Voltage Variable Capacitors**. The voltage variable capacitor was developed by Mucon Corporation. Varying the voltage of the capacitor could change its capacitance up to about 70 percent.

487. **Oxide Masking**. Bell Telephone Laboratories developed oxide masking techniques and diffusion for transistor manufacturing. This process of masking and diffusion led to major advances in the development of integrated circuits.

488. **Manufacturers**. By the middle of 1954, there were at least 70 foreign and domestic companies manufacturing transistors and diodes. The first transistors made in Japan were developed by Tokyo Telecommunications, later named Sony.

489. <u>Texas Instruments, Inc.</u> Texas Instruments, Inc. developed a method of growing silicon crystals with controlled impurity distribution. The silicon grown-junction transistor was produced by this method. Silicon provides a wider temperature-operating range for the transistors. The development was announced by Gordon Teal at the National Conference on Airborne Electronics. Texas Instruments soon became the major supplier of semiconductors to the computer industries.

490. <u>Relays.</u> Miniature relays, small enough to fit in a transistor, came on the market.

491. <u>Sylvania.</u> Sylvania introduced the stacked vacuum tube to compete with transistors. It was much smaller than the usual tube and operated at temperatures of about 195 to 500° C. The tube was never able to compete, however, because of the heating power required and the high temperatures resulting. Sylvania also put out the 1B3GT, 1X2B, 5U4GB, 6BQ6GTA, and 6SN7GTA.

492. <u>RCA.</u> RCA announced 20 new series string tubes for 600 ma heaters. These were: 2AF4A, 3AL5, 3AU6, 3AV6, 3BY6, 3CF6, 4BQ7, 4BZ7, 5AQ5, 5AS8, 5X8, 6AU7, 6S4A, 6SN7GTB, 12AX4, 12BH7A, 12BQ6GTB, 12BY7A, 12W6GT, and 25CD6GA.

COMPUTERS

493. <u>IBM 650.</u> The IBM 650 computer was made available this year. Over a thousand of these machines were sold. It had an 800 microsecond add time and a multiplying time up to 200 microseconds.

494. <u>Tradic.</u> The Bell Telephone Laboratories announced the development of a Transistor Digital Computer (Tradic). The computer contained about 800 transistors and required only about 100 watts of power.

495. <u>Univac.</u> The first Univac installation for business was at the G.E. Appliance Park, Louisville, Kentucky.

ILLUMINATIONS

496. <u>75,000-Watt Lamp</u>. The largest lamp in the world, 75,000 watts, was built to commemorate the 75th anniversary of the invention of the light bulb.

INDUSTRY

497. <u>The Tokyo Telecommunications Engineering Corporation</u>. The Tokyo Telecommunications Engineering Corporation produced the first successful output of transistors in Japan.

498. <u>William Shockley</u>. William Shockley left Bell Laboratories to start his own company--The Shockley Semiconductor Laboratory in Palo Alto, California. The company soon became a subsidiary of Beckman Instruments.

499. <u>RCA Victor</u>. The RCA Victor Division, which had been the manufacturing arm of RCA, was divided this year into two groups, RCA Consumer Products and RCA Electronics Products. A third group was organized at this time, RCA Sales and Service Subsidiaries. These changes were made to provide improved research, sales and manufacturing capability with greater diversity of products.

INSTRUMENTATION

500. <u>Hewlett-Packard</u>. Hewlett-Packard introduced the first high-frequency VTVM (vacuum tube voltmeter), the model-400D. The meter gave accurate indications at frequencies up to 4 MHz.

501. <u>Sextant</u>. Under contract with the US Navy, Collins Radio Company has developed a radio sextant for tracking the sun all day through heavy overcast with an accuracy better than could be obtained with the conventional sextant.

502. <u>400-Cycle Variac</u>. To meet requirements of the higher frequency power sources in use today, the General

Radio Company brought out a new line of Variac
auto transformers, ruggedized for marine and aircraft use. These were the M-2 and M-5 units designed for 350 to 1,200 Hz at 115 volts.

MASERS

503. J.P. Gordon, H.J. Zeigler and Charles Townes. Gordon, Zeigler and Townes developed the ammonia
maser at Columbia University. This was an amplifier which added negligible noise to the signal and
was used to amplify a 23,000-MHz signal. The term
"maser" stands for Microwave Amplification by Stimulated Emission of Radiation. This development later
led to the laser.

PHYSICS

504. Max Born and Walter Bothe. Max Born of England
and Walter Bothe of Germany were awarded the
Nobel Prize in physics--Born for his work in quantum mechanics and Bothe for work in cosmic radiation.

POWER

505. Atomic Energy Act. The Joint Committee on Atomic
Energy prepared a new atomic-energy law known as
the Atomic Energy Act of 1954. The act was signed
into law on August 30. This act modified the Atomic
Energy Act of 1946 to permit, among other things,
wider participation of private industry in nuclear
power development and industry, and private organizations to develop their own ideas of nuclear power.
Private ownership of nuclear reactors was allowed
with licensing by the Atomic Energy Commission.

506. Parallel-Wire Power Conductors. In Sweden, a
380,000-volt transmission line was built from the
power station at Harspranget to the south end of
Sweden, 600 miles away. This was the first time
parallel wires were used rather than one large con-

ductor to reduce power loss by air ionization and brush discharge.

507. Submarine Cable. A 100 kv submarine cable line was energized from mainland Sweden to the island of Gotland (60 miles). The system used direct current to provide 20,000 kw with mercury arc conversion to AC on the island.

508. Nuclear Power. The Russians put their first commercial nuclear power plant in operation in Obninsk. It was fully operational in June with 5,000-kw capacity.

509. Shippingport Reactor. On September 6, ground was broken for the construction of the first civilian-controlled nuclear power plant in this country. President Eisenhower started a process in Denver, Colorado that used a neutron source and fission detector to generate a signal transmitted across country to start a remote-controlled bulldozer in Shippingport, Pennsylvania to break ground for construction.

510. Block and Tabor. The idea of utilization of thermal energy stored in salt ponds for the generation of electric power was conceived about this time by Rudolph Block and Harry Tabor. The idea was put into practice about 25 years later.

RADIO

511. Pocket Radio. By December, the Regency division of Industrial Development Engineering Associates in Indianapolis, Indiana put on the market a four-transistor superheterodyne receiver, small enough to fit in a coat pocket. It sold for about $50 and was manufactured for Regency by Texas Instruments, Inc.

RADIO ASTRONOMY

512. Mills Cross Antenna. An antenna operating in the two-

to-four-meter region became fully operational at
Radio Physics Laboratory near Sydney, Australia in
August. The antenna was arranged in the shape of
a cross, formed by two 1,500-foot long dipole ar-
rays in a north-south and east-west direction. Each
array used two rows of 250 half-wave dipole elements over a wire-mesh reflector. The antenna was
reported to have a beam width of 49 minutes of arc.
The beam can be directed by the dipole phase adjustment.

SUPERCONDUCTIVITY

513. Superconductivity. It was about this time that the
     material niobium-3 tin was discovered to become
     superconducting at 18° K. This was the highest
     known temperature at which any known material be-
     comes superconductive and made the idea of super-
     conductive transmission lines, transformers and
     magnetic levitation for transportation come within
     the realm of feasibility.

TELEVISION

514. Color Television. Some color TV programs were being
     televised around the country. The first big event
     was the colorcast of the Tournament of Roses Parade
     in Pasadena, California on January 1. On March 17,
     RCA started a limited production of the color sets
     at its Bloomington, Indiana plant. These were 15"
     screens and sold for about $1,000. Before the year's
     end, colorcasting became a regular event in many
     cities.

1955

COMMUNICATION

515. Dr. John R. Pierce. The possibility and advantages
     of using satellites for communication was published

by Dr. John R. Pierce of the Bell Laboratories who specified such details as power, bandwidth and orbit parameters. As a result of this, the research department eventually set up Project Telstar.

516. <u>Tropospheric Scatter</u>. The first operational tropospheric scatter communication system was put in operation this year. This system provided a "gap-filler" at ranges not covered by UHF or SHF. It was known as Project Polevault.

COMPONENTS

517. <u>Mesa Transistors</u>. The diffusion process was applied to controlling the depth of penetration of impurities into germanium or silicon. The diffused base transistor provided a much-improved high-frequency performance. It was called the Mesa Transistor.

518. <u>Solar Cells</u>. Scientists at the Princeton Laboratories of RCA developed the gallium arsenide photovoltaic cell. The first GaAs cell was about one square millimeter in area with an efficiency of about six percent. In the next five years the efficiency was doubled.

519. <u>Varactors</u>. The voltage-sensitive capacitors known as varactors were developed at Bell Labs about this time.

COMPONENTS

520. <u>Tubes</u>. Since the advent of the transistor the number of new tubes was on the decline, except for transmitting tubes, with which the transistor could not compete. Many of the new tubes were improved versions of other models.

Eimac brought out some high power klystrons for high-frequency work up to 5,700 MHz with input power varying from 50 to 50,000 watts. Other tubes announced by Eimac included improved versions of 3X3000F1, 4X250B, 4X250F, 4CX5000A, and 4X250M. Penta Labs announced PL6549 and 6569; General

Electric brought out the 6B6GA and 6L6GB; and
Sylvania the 6W4GT, 6CS6, 6BN6. General Electric
also announced a new microminiature tube 3/8" long
and 1/2" in diameter. It operated in the UHF range
at temperatures over 500° C.

COMPUTERS

521. Erma. This year the General Electric Co. opened its
computer manufacturing facilities in Phoenix, Arizona
to produce an Electronic Recording Method of Ac-
counting (Erma).

522. Eniac. On October 2, the Eniac, built in 1946 at the
University of Pennsylvania, was switched off and
retired from service at the Aberdeen Proving Grounds
in Maryland.

523. Norc. The Naval Ordnance Research Calculator (Norc),
developed by IBM engineers, was delivered to the
Navy's Computation Laboratory. This machine was
capable of 15,000 calculations per second. Access
time from its cathode-ray tube memories was eight
microseconds.

524. Bizmac. The US Army Ordnance Tank-Automotive
Command in Detroit, Michigan installed a four-unit
RCA Bizmac to control the Army's inventory for
tank and automotive spare parts, which included
over 100 million parts. This was the world's largest
computer at that time.

CONFERENCES

525. Administrative Radio Conference. An Administrative
Radio Conference of the International Telecommunications Union convened August 17 to review and revise, if necessary, the world's radio communications
regulations which had resulted from the seventh International Radio Conference in 1947. About 80 nations were represented. Consideration of the frequency allocations for the various services was the
primary work of the conference. The new regulations were to go in effect as of May 1, 1961.

526. <u>Electrostatic Generator</u>. This year, two Van de Graaff generators were used to build up energy on the order of twice that of a single machine. With such a system, protons have been accelerated to an energy level of at least 25 million electron volts.

FABRICATION TECHNIQUES

527. <u>Fabrication Machine</u>. A machine was announced that could be programmed by punched-tape commands to wire electronic circuits. This was developed by the Bell Telephone Laboratories. The machine used the wire-wrapping techniques which made soldering unnecessary.

INDUSTRY

528. <u>RCA</u>. Two new operational units of RCA were set up with the split of the Electronics Product Division into the Defense Electronics Product and the Commercial Electronics Products divisions.

INSTRUMENTATION

529. <u>Microwave Oven</u>. The microwave oven was introduced by Tappan. A magnetron tube developed the energy to cook the food.

530. <u>E.W. Muller</u>. The field-ion microscope was invented by E.W. Muller of the Pennsylvania State University. The device was of particular value in studying the crystal lattice.

MASERS

531. <u>Bloembergen, Basov, and Prokorov</u>. Optical excitation for masers was proposed by Nicholas Bloembergen in the US and almost simultaneously by Nikolai Basov and Prokorov in the USSR.

PHYSICS

532. **R. Braunstein.** Infrared radiation was noticed by R.
Braunstein due to carrier injection in certain types
of semiconductors. This was a step in the direction
of the light-emitting diode produced a few years
later.

533. **Polykarp Kusch and Willis E. Lamb.** The Nobel Prize
in physics was awarded to P. Kusch and W.E.
Lamb, both of the United States, for their experiments in the quantum theory of radiation.

RADIO

534. **Car Radios.** Hybrid car radios came out this year, a
step towards the all-transistor sets.

RADIO ASTRONOMY

535. **B. Burke and K.L. Franklin.** B. Burke and K.L.
Franklin discovered radio noise from the planet
Jupiter on about 22 MHz. Later investigations
showed Jupiter to be a source of sporadic emissions
of radio frequency noise with durations lasting from
a few seconds to several hours and over a wide range
of frequencies.

SOUND

536. **Electronic Sound Synthesizer.** A public demonstration
of an electronic sound synthesizer was given by
RCA. The instrument was reported to simulate the
sound of musical instruments with remarkable fidelity.

STANDARDS

537. **Frequency Standards.** A cesium-beam frequency standard was put in operation at the National Physical
Laboratory in England as a time-standard control.

TELEPHONE

538. Newfoundland-Scotland Cable. Laying of the transatlantic telephone cable from Clarenville, Newfoundland to Oban, Scotland (2,000 NM) was started in June by AT&T and the British and Canadian communications agencies. This was a coaxial cable with a vacuum tube amplifier every 40 miles. Cable was laid by the HMTS Monarch.

539. Solar Power. The Southern Bell Telephone Co. used solar-generated power for the first time to power a commercial telephone conversation on October 4.

TELEVISION

540. 21" Color Sets. The first 21" color sets by RCA were marketed in January. The picture tubes were round at this time.

541. British T.V. Commercial TV started in England on September 22.

542. Color City. NBC opened "Color City" at Burbank, California. It was the first studio designed especially for color TV. By the end of the year at least 111 NBC stations were telecasting in color.

543. Alabama Educational TV. The first state educational TV network was established in Alabama, April 28.

544. Detroit Area. Closed-circuit television surveillance for traffic control was instituted in the Detroit area.

WHISTLERS

545. Morgan and Allcock. The theory of whistlers was being studied by Millett G. Morgan and G. McK. Allcock. They used synchronized recorders in New Zealand and in the Aleutian Islands on opposite ends of a magnetic line of force of the earth's magnetic field. On August 28, when lightning struck in New Zealand, 1 1/2 seconds later a whistler was heard at the north end of the magnetic line.

<u>1956</u>

AMATEUR

546. <u>50 MHz DX Record</u>. A new record for distance on the
50-MHz band was reported to have occurred on
March 24, when LU9MA in Mendoza, Argentina worked
JA6FR on Kynshu Island, Japan. The distance was
nearly 12,000 miles.

COMPONENTS

547. <u>Miniature Transistors</u>. The trend towards miniaturization did not end with the development of the transistor. By the middle of the year Philco was producing PNP transistors small enough that 20 could
be placed on a dime. The transistor had a 70-db
gain with a power gain of about ten million.

548. <u>Solid-State Switches</u>. The first solid-state silicon
switches were put on the market by General Electric
Co. this year.

549. <u>Tubes</u>. Relatively few new tubes were announced this
year. Some of the new tubes announced by their
manufacturers included: Penta Labs--PL172 and
Eimac--4CX250K, 4CX300A and the X602 Klystron.
This was a six-foot-high tube producing 75 kw continuous wave output in the UHF region. Eimac also
made ceramic tubes to be tested by the Air Force.
Characteristics were similar to those of the 6SN7 and
6AK5 tubes.

550. <u>Tube Replacement Started</u>. The change from vacuum
tubes to transistors for military equipment started
this year. Military computers required transistors
for reliable operation and reduced power requirements.

FABRICATION

551. <u>Thermocompression Bonding</u>. The thermocompression

bonding technique of attaching electrical leads to
semiconductor devices was developed about this time.
The resulting bond was stronger than the attached
lead.

552. Frequency Standards. The first commercially available
cesium-beam frequency standards appeared this year.
The development was headed up by R. Daly of the
National Co. and was known as the Automichron.
The accuracy was improved one order of magnitude
about every five years up to at least 1970.

INSTRUMENTATION

553. Oscilloscopes. Hewlett-Packard came into the oscilloscope market with the model 130A (300 MHz) and the
150A (10 MHz) versions.

554. Maser. Solid-state microwave amplification by stimulated emission of radiation was demonstrated by Bell
Labs in November. The maser soon became the
standard where microwave low-noise receivers were
required as in radio astronomy and satellite communication.

MEDICAL

555. Medical Electronics. The depth of anesthesia during
surgery could be monitored by a new device which
made heart and brain-wave tracings. Danger signals
not otherwise detectable could be seen with increased
reliability and freedom from the otherwise possible
brain damage.

556. Paul M. Zoll. Dr. Paul M. Zoll of the Beth Israel Hos-
pital in Boston announced the pacemaker for electrical
stimulation of the heart. The device was attached
externally at this time.

PHYSICS

557. US Air Research and Development Command. In an ex-

periment of discharging a high current through a high-pressure gas, scientists at the US Air Research and Development Command, Cambridge, Massachusetts are reported to have produced a flash roughly 700 times the brightness of the sun and an instantaneous temperature of 400,000° F.

558. **William Shockley, John Bardeen and Walter H. Brattain.** William Shockley, John Bardeen and Walter H. Brattain of the US shared the Nobel Prize in physics for invention of the transistor.

POWER

559. **Atomic Power.** England's first atomic power plant went into operation at Clader Hall. The plant generated 90,000 kw of power.

560. **Turbo Generators.** Steam turbine generator units capable of developing 450 MW were developed in the United States.

561. **Solar Power.** A unique radio was announced this year. Although it required a battery for use at night, solar cells provided the power to keep the batteries charged and to power the seven transistors on bright days.

RADAR

562. **DEW Line.** The idea of an early warning alarm system for an over-the-pole missile attack was conceived in 1952 and construction was started in 1954. Before being fully completed, parts of the Distance Early Warning Line (DEW Line) were put in operation October 24. Contract for its operation and maintenance was awarded to the Federal Electric Corporation.

RADIO

563. **NAA Closed Down.** After 43 years of service, the first high-power naval station, Radio Arlington--NAA--was

closed down. Due to the 600-foot towers becoming
a hazard to air traffic in the Washington area, the
towers were taken down. The deactivation cere-
monies were held at Radio Arlington on July 14.
The principal speaker was Rear Admiral H.C. Bru-
ton, W4IH.

RADIO ASTRONOMY

564. J.D. Kraus. Radio outbursts from Venus on 26.7
MHz were observed this year by J.D. Kraus. These
were short bursts lasting from less than a second
to several seconds with bandwidth of at least 2
MHz.

565. Radio Astronomy Zoning Act. The Radio Astronomy
Zoning Act was passed in West Virginia on August
9. This act restricted licensed electrical equipment
generating interference above certain levels from
being operated within ten miles of the Green Bank
Radio Astronomy Laboratory.

566. R. Coutrez, J. Hunaerts and A. Koeckelenbergh.
What is believed to be the first radio emission de-
tected from a comet was observed on 600 MHz by
Coutrez, Hunaerts, and Koeckelenbergh of the Royal
Observatory of Belgium in Uccle, Belgium.

TELEPHONE

567. Cables. The first submarine telephone cable service
between the United States and Great Britain was put
in operation by the Bell System. This was a 36-
voice channel. It was later increased to 96 channels
using time assignment speech intermodulation (TASI).
The cable joined New York, Montreal and London.
Service started September 25.

<u>1957</u>

AMATEUR

568. <u>144-MHz DX Record</u>. The world's record for DX on 144 MHz was broken on July 8, when KH6UK made contact with W6NLZ on 144 MHz. This range of 2,450 miles was about twice the distance record set in 1950 which had held up to this time.

569. <u>Project Moonbeam</u>. For about a year and a half the Naval Research Laboratory had been working on a system of satellite tracking called Minitrack. An invitation was issued to qualified amateur groups to assist in this effort of satellite tracking. The program, known as Project Moonbeam, was expected to provide information to permit a more exact determination of the size and shape of the earth. The amateurs were expected to provide valuable data for the project.

BROADCASTING

570. <u>Broadcasting</u>. About this time, interest in FM broadcasting began to increase, partly because of improved receivers, automatic frequency control, the interest in high-fidelity music, and the possibility of stereo reproduction. The demand for FM licenses increased. Many organizations requested licenses, just to have them, before the band was filled.

COMMUNICATION

571. <u>Harold F. Meyer and Walter E. Morrow</u>. At a meeting to discuss the possibility of using satellites to extend the range of microwave communications to intercontinental ranges, the idea of using resonant dipoles was suggested by Harold F. Meyer and Walter E. Morrow. For command communication they could not depend on the reliability of an orbiting satellite and preferred all active electronic equipment to be on the ground. This is the beginning of what later became project West Ford.

COMPONENTS

572. <u>W.T. Read, Jr. and Leo Esaki</u>. The theory of the Impact Avalanche Transit Time (IMPATT) diodes was conceived by W.T. Read, Jr. of the Bell System and was described the following year in the Bell System Technical Journal. The idea was further developed by Leo Esaki of the Sony Corporation who discovered the tunneling effect.

573. <u>Medical</u>. A microphone, 0.05 inches in diameter and small enough to be passed through a vein or artery into the heart, was designed to study heart sounds without the noises caused by breathing or the digestive tract activity.

574. <u>Tubes</u>. Fewer and fewer new tube types were being announced as transistors began taking over in many applications. Transmitting tubes were not yet being displaced except in some lower-power applications. New types were brought out by at least three companies: Amperex--5893; Eimac--4CX1000A; RCA--7034/4X150A; and Burroughs Corp.--Nixie (a numeral readout tube).

575. <u>Silicon Transistors</u>. The first silicon power transistors were developed this year. Yields were low and the price was high. These were NPN types, rated to up 300 v.

576. <u>Gate-Controlled Rectifiers</u>. The gate-controlled silicon rectifier, having a gate electrode to control the firing point, was developed by the General Electric Laboratory. These are generally called thyristors or silicon-controlled rectifiers.

COMPUTERS

577. <u>Memory Systems</u>. Improved memory systems were being developed. IBM developed the RAMAC rotating disc memory, a random access device capable of storing up to five million characters. During this period, Bell Labs developed the plated wire memory.

INDUSTRY

578. Grinich, Last, Hoerni et al. Fairchild Semiconductor
Company of Mountain View, California, was founded
in September by Victor Grinich, Jay Last, John
Hoerni, Eugene Kleiner, Julius Blank, Gordon Moore,
Robert Noyce and Sheldon Roberts. Noyce became
director of R & D and the company soon rose to become one of the leaders in semiconductor development.

579. Olsen and Anderson. The Digital Equipment Corp.
was organized in August by Ken and Stan Olsen and
Harlan Anderson, to manufacture digital logic mod-
ules. They produced the programmed Data Processor
(PDF-1) in 1959.

580. Sony Corporation. In January, the name Sony was
adopted as the name of the firm previously known
in Japan as Tokyo Tsushin Kogyo Kabushiki Kaisha
(in English, Tokyo Telecommunications).

INSTRUMENTATION

581. Memo-Scope. The Memo-scope was announced by
Hughes Aircraft. This scope retained the traces
from single-sweep operation indefinitely or until
erased.

582. Micro-microammeter. In July, Keithley Instruments,
Inc. announced the first instrument with high stability, capable of measurements in the region of
10^{-3} to 10^{-11} amps. It was used in measuring in-
sulator leakage, capacitor testing, currents in ion
chambers and photocells.

583. Capsule Transmitter. A small FM transmitter in a capsule was developed by RCA to permit the monitoring
of body functions while the person remained active.

LASERS

584. R. Gordon Gould. The term "laser" (light amplification

by stimulated emission of radiation) was attributed
to R. Gordon Gould while at Columbia University.
Gould was awarded a patent on an optically pumped
laser amplifier.

PATENTS

585. John T. Wallmark. John T. Wallmark of the RCA Laboratories was granted a patent on the field-effect
transistor. He did not develop an operating model,
however.

POWER

586. Atomic Power. The world's first full-scale, civilian-
operated nuclear power plant in the US was put in
service on December 18. This was the Shippingport
reactor, located about 25 miles from Pittsburgh, on
the Ohio River. Development was financed by the
Atomic Energy Commission.

587. Japanese Atomic Power. In September, the first atomic
reactor in Japan began producing power in Tokai-
mura.

588. Gas-Cooled Reactor. The development of the High-
Temperature Gas-Cooled Reactor (HTGR) was started
by GA Technologies of San Diego, California. The
HTGR system was considered safer, with essentially
zero radiation leakage to the environment. The
coolant used was helium, an inert gas that does not
become radioactive.

RADAR

589. Monopulse Radar. The first monopulse tracking radar,
the XN-1, was installed at Patrick Air Force Base
this year.

590. DEW Line. The Distant Early Warning Line (DEW Line)
radar system, started in 1955, was completed and
put into operation. The line extends across Northern

Canada from Cape Lisburne on the Alaska Coast to Baffin Island on the East (approximately along the 70th parallel). The DEW Line was built to detect any attack on the US or Canada from the USSR.

RADIO

591. <u>World Spanner Radio</u>. The Army announced that it would install a 24-million-watt "World Spanner" short-wave transmitter capable of being heard anyplace on earth. The transmitter was designed at the US Army Signal Engineering Laboratories and the Continental Electronics Mfg. Co.

RADIO ASTRONOMY

592. <u>Jodrell Bank</u>. The Radiotelescope at the Jodrell Bank Experimental Station was put in operation in October of this year.

593. <u>National Radio Astronomy Observatory</u>. Construction of the NRA Observatory was started this year near Green Bank, West Virginia. The site required 2,700 acres and was at an elevation 2,700 feet above sea level. At this time the largest radio telescope in the United States was the 60-foot dish at Harvard University, followed by the 50-foot dish at the Naval Research Laboratory near Washington, DC.

RECORDING

594. <u>Stereophonic Recording</u>. A great advance in the recording and reproduction of sound was made with the introduction of stereophonic recording systems. The recordings were made with two microphones and the playback was made through two speakers. The right- and left-hand speakers presented an impression on the listener much as though he/she were sitting before the orchestra or artists. Later the 45/45 system of recording was announced and the Standards Committee of the Recording Industry Association of America adopted the system, which was

considered superior to the vertical-lateral recording
system previously used.

SATELLITES

595. <u>Satellites</u>. The first man-made satellite, <u>Sputnik 1</u>,
was put in orbit on October 4 by the Soviet Union.
It weighed 184 pounds at launch and was put in a
90-minute orbit. This was the beginning of the
Space Age.

SUPERCONDUCTIVITY

596. <u>Bardeen, Cooper and Schrieffer</u>. A theoretical explanation of superconductivity known as the "BCS"
theory was brought out by John Bardeen, Leon
Cooper and Robert Schrieffer.

TELEPHONE

597. <u>Tropospheric Scatter Communication</u>. On September
12, a telephone link from the US to Cuba was put
in operation using the tropospheric scatter mode of
operation. The system can handle two TV channels
each way or several hundred telephone calls. The
first tropospheric scatter system in the Netherlands
was set up between Denhelder and Domburg.

598. <u>New Telephone Service</u>. Telephones mounted by the
curb had been provided in some cities on a trial
basis. A driver could drive up to the phone and
talk without having to get out of the car. Also,
pocket pagers used to call people to a telephone
were put in service on an experimental basis.

599. <u>Telephones</u>. By the end of the year, telephones in
the US could be connected to over 96 percent of all
the world's telephones. Approximately 87 percent of
the telephones in the US were dial-operated.

TIME

600. <u>Atomic Clock</u>. An atomic clock, built at the International Telephone and Telegraph Laboratories, had been developed with an accuracy on the order of one second in a century.

WHISTLERS

601. <u>R.A. Helliwell and E. Gehrels</u>. In January, Gehrels and Helliwell of Stanford University, observed artificial signals propagated by the whistler mode. The signals of NSS in Annapolis on 15.5 kHz propagated along lines of force of the earth were heard at Cape Horn, South America.

<u>1958</u>

AMATEUR

602. <u>1,296-MHz Record</u>. On July 20, a new record for 1,296 MHz was made over a 225-mile path by W6MMU/6 and W6DQJ/6. This record was again broken in September by W6MMU/6 and K6AXN/6 in operating over a 270-mile path.

603. <u>Narrow-Band Picture Transmission</u>. A method of narrow-band picture transmission based on the Bell Telephone Laboratories Picturephone system was described in the August issue of <u>QST</u>. Using the image-storing cathode-ray tube (Iatron), a 60-line picture could be sent over ordinary phone lines. The all-electronic system required no more bandwidth than conventional amplitude-modulated phone signals. The system was first used at the amateur station of Copthorne Macdonald W4ZII/2.

604. <u>Call-Letter Shortage</u>. In spite of the call-sign system revision set up in the 40's, the need for call letters had so increased that a revision was again needed. This was first evident in the second and sixth dis-

trict. The new system would be used when W and
K prefixes ran out. New calls would be issued with
WA as a prefix. Novice licenses would be issued
with a WV prefix and be changed to WA when the
license was upgraded. WN prefixes would be issued
to novices in areas where the supply of calls was
sufficient.

605. WAS Award. The first Worked-All-States (WAS) award
for radio-printer contacts was made to Boyd Phelps
(WØBP) for the achievement. Verification was provided by 100 percent QSLs. A certificate was issued on May 2.

COMMUNICATION

606. Dataphone. The Dataphone concept was developed by
the Bell Laboratories and introduced this year. The
Dataphone permitted interchangeable use of telephone
lines for voice or data transmission. A general up-
dating of telephone lines to handle data was started
about this time.

607. Citizens Band. In September the FCC opened the
Class D Citizens Band at 27 MHz. No license was
required.

608. Project SCORE. In July, 1958, Project SCORE, the
acronym for Signal Communication by Orbiting Relay
Experiment, was approved and assigned to the US
Army Signal Research and Development Laboratory.
Ground stations were to be in operation by November
1. The system was devised by the military after
considering the vulnerability of line-of-sight microwave systems to earthquakes or military action. Two
modes of operation were to be demonstrated. The
delayed repeater, where a recorded message could
be delayed and played back over any part of the
earth, and a real-time repeater mode. The system
was launched into orbit on November 18 and served
to demonstrate both voice and teletype relay over
intercontinental distances before its failure due to a
battery problem on December 30.

609. West Ford Project. In the spring of the year, the US Army Signal Corps asked MIT to study the possibility of transcontinental communication in the UHF regions by scatter from orbiting dipoles. This was the first major step in the West Ford Project.

610. Telegraph. By this time, the US had direct telegraphic service to 84 foreign countries.

611. Telex. The first customer-to-customer teleprinter exchange service between the US and Canada was completed this year by Western Union and the Canadian National and Canadian Pacific Telegraph Companies. The Telex system provided 24-hour service throughout Canada.

COMPONENTS

612. Batteries. A rechargeable lead-silver dry cell was developed at the US Naval Ordnance Laboratory. The cell was rated at 1.12 ± 0.03 v with a capacity of 1,500 milliampere hours. It could be stored uncharged with negligible degradation.

613. Mesa Transistor. Motorola Inc. announced their first two Mesa Transistors, the 2N695 and the 2N700. The 2N695 was a fast-switching transistor. The 2N700 was an amplifier operating at up to 200 MHz. These transistors were some of the first manufactured with tolerances so close that specially picked units were unnecessary. Mesa transistors capable of operating in the gigahertz region were developed by Fairchild.

614. Stanislas Teszner. Stanislas Teszner of Poland made the first Junction Field-Effect transistor (JFET) which he called the Tecnitron. The JFETs were produced in this country first by Crystalonics in Cambridge, Massachusetts in 1960.

COMPUTERS

615. Computer Printer. The Stromberg Carlson Division of

General Dynamics Corp. and the Haloid Co announced the development of an electronic printer for computers which could print out data at 4,680 lines per minute. This was an estimated five to ten times faster than the electromechanical type.

616. Philco. Philco announced its Model 2000 all-purpose data processing system. This was a high-speed machine using surface barrier transistors. Addition required 1.7 μs, multiplication, 40.3 μs. The memory required approximately 10 μs access time.

617. New Computers. Other second-generation computers came out about this time. IBM's 7090 had a 2.18 μs access time. Control Data Corporation brought out its 1604, a lower-priced machine, resulting in its popularity.

HEARING AIDS

618. Zenith Solaris. In August, Zenith developed the Solaris hearing-aid unit. The device was powered by solar cells and was designed to be mounted on the temple bar of eyeglasses.

INDUSTRY

619. Shockley Transistor Labs. The Shockley Semiconductor Laboratory changed its name to Shockley Transistor Laboratories with the intention of becoming more manufacturing centered.

620. General Telephone and Electronics Corporation. The Sylvania Electric Company organized in 1901, merged with the General Telephone Corporation and was reorganized as the General Telephone and Electronics Corporation.

621. Thompson-Ramo-Wooldridge (TRW). Ramo-Wooldridge, and The Thompson Products Company joined to form Thompson-Ramo-Wooldridge.

INSTRUMENTATION

622. Transistorized Flash Unit. The West German firm of
Metz Apparate Fabrik, announced its Megablitz-100
flash unit. These were reported to be the first
transistorized flash units developed.

623. Automatic Balancing Bridge. The Barnes Development
Company developed an automatic balancing capacitance bridge which indicated the capacitance and
dissipation factor of the condenser under test in an
average time of seven seconds per capacitor in the
range of 100 $\mu\mu$F to 1.1 μF, with dissipation factors
of from 0 to 16 percent.

INTEGRATED CIRCUITS

624. J.S. Kilby. J.S. Kilby of Texas Instruments developed
the phase shift oscillator from a single silicon bar.
Following the oscillator, he developed the IC flip-
flop circuit which was ready the next year. By the
end of the year, J.S. Kilby had built capacitors of
oxide layers on silicon and diffused layer resistors,
further advancing the science of integrated circuits.

MEDICAL

625. H.O. Anger. H.O. Anger of the University of California developed the gamma-ray camera. The camera
provided information on the biochemical activities of
the body, metabolic rates, location of tumors, and
other functions without the necessity of exploratory
surgery.

ORGANIZATIONS

626. National Association of Broadcasters. The National Association of Radio and Television Broadcasters
changed its name back to the original form--The National Association of Broadcasters.

POWER

627. Nuclear Power. The second Russian nuclear reactor was built in 1958 at Troitsk.

628. Thermionic Conversion. The direct conversion of heat to electricity was obtained at the General Electric Research Laboratory with an efficiency of about 8 percent. While this wasn't high efficiency it was greater than could be obtained from thermocouple types of converters.

RADAR

629. Ballistic Missile Early Warning System (BMEWS). As problems with the Soviet Union continued, the Department of Defense awarded one of the largest contracts it had ever let to RCA to provide early warning of a missile attack from over the pole. With this contract, RCA took project management of the Ballistic Missile Early Warning System. Nearly three thousand companies furnished material and expertise in completing the project.

RADIO ASTRONOMY

630. 1,000-Foot Dish Proposal. It was proposed this year to make a radio telescope with a 1,000-foot diameter dish by scooping out a valley near Arecibo, Puerto Rico. The proposal was made by Professor William E. Gordon of Cornell University.

RECORDING

631. Stereo Recording. Stereo records and pickups appeared on the market; the 45/45 record-cut made stereophonic audio reproduction practical for home use.

632. VERA. This year, the British Broadcasting Corp. used in its programming a recorder known as the Vision Electronic Recording Apparatus (VERA).

This machine operated at a tape speed of 200 ips
and gave satisfactory operation. It was eventually
made obsolete by the videotape unit developed by
the Ampex Corporation.

SATELLITES

632a. Project SCORE. The Signal Communication by Orbiting
Relay Experiment (SCORE) was a joint effort of the
Advanced Research Projects Agency, the Army and
the Air Force. The initial goal of the project was
to place in orbit an 80-foot-long Atlas Missile which
would carry a communication system, control unit
recorder and tracking beacon. This experiment
would give a good indication of the problems to be
expected in a satellite system. The Atlas Missile was
launched on December 18. The possibility of such a
satellite was demonstrated until battery failure on
December 30. This is considered the first message
by voice ever received from space.

SPACE

633. National Aeronautics and Space Act. (Space Act).
The Space Act of 1958 provided for a civilian agency
called the National Aeronautics and Space Administration generally known as NASA. Also, a National
Aeronautics and Space Council, presided over by
the President of the United States, was created to
control NASA. The function of NASA is to conduct
the aeronautics and space activities of the government except for weapons systems and military projects.

634. Project TRAC(E). What is believed to be the first
system for Tracking and Communication, Extra-
terrestrial--TRAC(E)--was developed under the US
Army Lunar Program, initiated in 1958 by the Jet
Propulsion Laboratory. Previous systems for tracking satellites were unsatisfactory at lunar distances
or beyond. The system developed was suitable for
simultaneous tracking of and communication with
lunar or space probes. A frequency of 960.05 MHz

was found to be optimum considering the state of the art at this time.

635- James Van Allen. The Van Allen radiation belt was
6.   discovered by James Van Allen of the University of Iowa. Data from the Explorer 1 flight showed unexplained gaps in radiation around the earth. Explorer 2 failed. Explorer 3 carried a radiation recorder when launched on March 25. Analysis of the recording indicated that the blackout of the radiation count was due to saturation of the counter, roughly 100 times that which was expected. On May 1, he announced that the earth was surrounded by a band of charged particles trapped by the magnetic field. This came to be known as the Van Allen Belt.

TELEPHONE

637. California-Hawaii Telephone. The first undersea telephone cable between California and Hawaii was completed this year by the British cable ship Monarch. The cable contained two coaxial cables transmitting in opposite directions. Amplifiers were built in at 40-mile intervals. An average life of 20 years was expected for the amplifiers. Each cable carried 36 simultaneous conversations.

638. Bell Telephone System. The Bell Telephone System had expanded its microwave system to where approximately one-quarter of its long-distance calls were handled by this means.

TELEVISION

639. Color TV. The Tuscany Hotel in New York City is believed to have been the first hotel with a color TV in every room.

640. Cable TV Systems. By this year, there were over 700 cable TV systems in the United States, primarily in mountainous areas but extending also to areas where only one station could be received.

TROPOSPHERIC SCATTER

641. <u>Tropospheric Scatter</u>. The first tropospheric-scatter system in France was installed for communication between Lannion and Conches.

Other tropospheric scatter systems were put in operation in Venezuela, Iceland, Norway and Algeria.

<u>1959</u>

AMATEUR

642. <u>Radioteletype</u>. Radio teletype communication for amateurs began in Great Britain this year.

643. <u>Amateur Television</u>. In a series of tests during November and December, amateur transatlantic television was accomplished. This transmission, by WA2BCW in Elmira, New York, was received at G3AST in Yeovil, Somerset ... a new first for the amateurs.

644. <u>D.E. Isbell</u>. The log periodic antenna was introduced by D.E. Isbell. The antenna has a very wide bandwidth, limited by size and construction details. In spite of its wide frequency response, gains of over 9 db were attained.

COMMUNICATIONS

645. <u>Tropospheric-Scatter Communication</u>. This year, Tropo Systems were put in operation in Libya, West Germany, Kwajelein and Eniwetok.

COMPONENTS

646. <u>Printed-Circuit Motors</u>. The printed-circuit motor was developed at this time for a low inertia motor to

speed up tape drives in electronic data processing equipment.

647. Jean Hoerni. Using diffusion and oxide masking techniques, Jean Hoerni of Fairchild Semiconductor Corporation developed the Planar Process of transistor manufacture. The Planar Process supplanted previous structures at Fairchild because of the complete oxide protection of all junctions. The protection against contamination resulted in improved electrical characteristics throughout the life of the device.

648. Tunnel Diodes. RCA developed tunnel diodes capable of operating at frequencies in the gigahertz (GHz) region. Interest in the application of the diode seemed to decrease in spite of improved diodes and reduced prices.

649. Paul Weimer. Paul Weimer of RCA built the first thin-film field-effect transistors in this country.

650. Atalla and Kahng. The oxide insulated-gate field-effect transistor was developed at the Bell Laboratories this year by M.M. Atalla and D. Kahng. It soon became the most widely used of the FETs.

651. Magnetron Tubes. Early in the year, the first hydraulically tuned magnetron was announced by Litton Industries and was known as the L-3211. It was the fastest tuning medium power magnetron up to that time. Operation was in the X-band.

652. Nuvistor Tubes. Nuvistor tubes came out this year, introduced by RCA. This was a metal porcelain tube, not a great deal larger than the transistors of the time. These tubes could not compete with the transistor and were not widely used. RCA also brought out the 7360 Beam Deflection tube developed for Single and Double Side-Band Systems.

COMPUTERS

653. DaSpan. A computer-to-computer communication system called DaSpan was introduced by the Radio Corpora-

tion of America. This system permitted data collecting from many points. Then data processing or other functions could be done remotely with more effective use of computer operating time.

654. ALGOL. The Algorithmic language (ALGOL) was developed by the Association for Computing Machinery and the German Association for Applied Mathematics. This was to be a common language which could be used by many computers.

655. Grace Hopper. Another attempt at a common computer language was the Common Business-Oriented Language (COBOL) developed at the Pentagon by a group under Captain Grace Hopper.

656. Automatic Sequence Controlled Calculator (Mark I). The first fully automatic computer generally known as Mark I was built at Harvard University and completed in 1944. It was put in operation at Harvard and used until 1959 when it was retired.

657. Whirlwind I. After nearly a decade of service, Whirlwind I was shut down and retired from service in May.

CONFERENCES

658. National Stereophonic Radio Committee. The National Stereophonic Radio Committee was set up to submit recommendations for FM stereo broadcasting.

659. Administrative Radio Conference. The Administrative Radio Conference of the International Telecommunications Union (ITU) met in Geneva, Switzerland on August 17 to consider and revise as necessary the world's radio communication regulations which had been adopted at the Atlantic City meeting in 1947. The meeting, represented by 87 member countries of the ITU, was called because the majority of the member nations had problems which needed consideration. The primary technical changes were made in the existing international frequency assignments.

INDUSTRY

660. Compagnie Française Thomson-Houston (CFTH). In 1928, CFTH had taken over the Ducretet Establishments, giving them an entrance into the radio and electronic fields. With an agreement with Pathé Marconi, by 1959 CFTH had become one of the leading manufacturers of radio and television equipment in France.

661. Hewlett-Packard Company. Since setting up operations in 1938, the Hewlett-Packard Company had seen a period of almost continual growth. In 1956 it had started a manufacturing complex in the Stanford Industrial Park and began to expand into even broader fields. By 1959 the company was expanding to various sites around the world--a marketing organization in Switzerland, a manufacturing plant in West Germany and still further expansion in the 1960's.

662. Norman Kjeldsen. In January, the Cardwell Manufacturing Company was purchased by Norman Kjeldsen. The company continued to expand and take over more companies and by 1981 Cardwell was manufacturing over 15,000 items.

INSTRUMENTATION

663. Transistorized Counter. The first transistorized counter, the Model 5310, was produced by Berkley Division of Beckman.

664. Robert Sugarman. An oscilloscope capable of displaying repetitive frequencies in the gigahertz region was developed at the Brookhaven National Laboratories by Dr. Robert Sugarman.

INTEGRATED CIRCUITS

665. J.S. Kilby. The first integrated circuit flip-flop was introduced by J.S. Kilby of Texas Instruments at the Institute of Radio Engineers Show.

666. Robert Noyce. Robert Noyce of Fairchild conceived the idea of diffused resistors and interconnections by evaporated metal. The interconnecting idea of Noyce was soon adopted by Texas Instruments.

POWER

667. The SNAP Generator. The System for Nucear Auxiliary Power (SNAP) was announced on January 16. SNAP was a thermoelectric generator operating on a radioactive isotope which generated power by the heat of radioactive decay operating on a series of thermocouple elements. These elements are generally semiconductive materials for more efficient power conversion. The first model could deliver a total of about 2.9 kwhr over a 280-day period.

668. Solar Cells. Silicon Photovoltaic cells (solar batteries) were developed by Bell Telephone Laboratories. By this date, they had been used to power satellites giving a continuous source of power to Vanguard I, Sputnik III and Explorer VI and VII. In outer space these solar cells theoretically operated at ten percent efficiency and 1,400 watts per square meter. In actual practice, considering orbit, spinning, temperature, cell protection, dust, etc., actual power averages something over 7 watts per square foot.

POWER CONTROL

669. W.E. Campbell. An article in the June 8 edition of Electrical World by W.E. Campbell, describes a switch capable of controlling a three-phase 230 kv, 600A transmission line in Arizona. Eight vacuum switches were used in the circuit.

RADAR

670. Missile Early Warning System. The first Ballistic Missile Early Warning System Station (BMEWS) was assembled in Greenland this year. The transmitter had

a peak power of over 2 Mw and could theoretically
detect a missile at 3,000 miles.

RADIO

671. Pioneer 4 Lunar Probe. The first application of the
equipment developed under the TRAC(E) program
was put on the Pioneer 4, launched March 3. Tracking was good to about 407,000 miles when decreasing
battery voltage resulted in a rapid decrease in signal strength. The 407,000 miles were a new record
for radio.

RADIO ASTRONOMY

672. Philip Morrison and Gioseppe Cocconi. After hearing
the noise from outer space picked up by radio telescopes, it occurred to Morrison and Cocconi that
if there were intelligent life forms on any other
planets in the universe, there was a possibility that
their signals, if any, might be picked up on the
radio telescopes.

673. Frank Drake. The idea of extraterrestrial intelligence
had occurred to Frank Drake at the National Radio
Astronomy Observatory in Green Bank, West Virginia
at about the same time as it occurred to Morrison
and Cocconi. He was planning a search with the
85-foot observatory telescope.

REFRIGERATION

674. Thermoelectric Cooling. This year, Westinghouse and
Whirlpool demonstrated refrigeration using thermoelectric techniques.

SATELLITES

675. Vanguard 2. The first of the weather satellites to
give reliable meteorological returns from space was
Vanguard 2. The returns from satellites permitted
a breakthrough in long-range weather forecasting.

TELEPHONE

676. <u>Undersea Cable</u>. Another deep-sea telephone cable was put in service connecting New York and Montreal with Paris, France and Frankfurt-am-Main.

TELEVISION

677. <u>Transistorized TV</u>. Japan's transistorized television sets were introduced in the US by the Sony Corporation. The original lines were black-and-white sets with color soon to follow.

678. <u>India</u>. Television was introduced in India this year and could be seen in six community centers in New Delhi. TV stations would be built in about ten Indian cities within five to six years.

679. <u>Cuba</u>. Color television broadcasts were started in Cuba.

Chapter 3

THE SEVENTH DECADE: 1960-1969

The decade of the 1960's brought a number of advances in communications, via cable, radio, satellite and troposcatter Television kinescopes and sets were being improved in brightness and color reproduction. Computers were improved in speed and reduced in size and power consumption, electron microscopes provided unprecedented magnification, lasers developed into military weapons. In the power field, atomic power became economically competitive with fossil fuel plants. The radio astronomers were making breakthroughs in understanding the universe. Radar reflections from many of the planets were detected. All in all, it was a fantastic decade for the electrical and electronic fields.

The decade opened with experimental transcontinental communication by a passive satellite. After a period of many relay satellite launches, by the end of the decade satellite relay communication with points all over the world became a common event. Live television from Europe and Japan were accomplished and satellite relay became an accepted means of communication. During this period the longest telephone call on record was made by the President of the United States from the White House to men on the moon.

In addition to the communication satellites in orbit, a number of other satellites were launched. Satellites for atmospheric studies, monitoring for illegal atomic tests, weather prediction and other purposes were put in orbit. Space probes were launched and many thousands of pictures were received from space and the surface of the moon. Some of these space probes provided valuable data on magnetic fields in space, solar wind, ionization and other information. Two orbiting solar observatories were launched to monitor the effect of solar phenomena on the terrestrial and interplanetary environment. Two orbiting geophysical observatories were also launched specifically to study the rays, protons, electrons, solar wind and very low frequency propagation.

The expansion of telephone lines to remote parts of the earth continued. A number of new submarine cables were installed, joining the United States to a number of new areas in the Pacific such as Australia, Japan, Midway, Wake and Guam. More cables to Europe provided many additional voice channels. These additional channels were required because of the increasing number of transatlantic calls being made each year. In 1960 alone, about three million transatlantic telephone calls were completed. In the US, the telephone system was continually being improved by the installation of high-speed electronic switching and methods of increasing cable capacity. By the end of the decade, electronic switching system conversions had about been completed in the United States and the system was being introduced in some foreign countries. Telephone lines in this country were being hardened against earthquakes or atomic attack. During this period, the 100-millionth telephone was installed.

Tropospheric-scatter telephone systems were put in service in many parts of the world from the Azores to many areas in Asia and the Pacific Islands. Microwave systems were also being installed. Western Union began operation on a transcontinental microwave system increasing their capabilities in both channel-miles and words-per-minute transmission. The Canadians installed a microwave system which was considered the world's most advanced up to that time. This system provided both telephone and telegraph channels.

There were some experiments with four-channel stereophonic reproduction but little came of it. The idea was that having two additional speakers to simulate the normal reflection from the rear of the auditorium would lend reality to the reproduction. Two stereo stations cooperated in this experiment but after the initial announcement little more was heard of four-channel stereo for some time. Television was another story. With the opening of the decade, a new color-tube for the television cameras was announced. The color pickup tube required no more light than was required for black and white cameras. Near the end of the decade, camera tubes requiring still less light came into use. Improved picture tubes for receivers gave better color rendition, sharper definition, and greater brightness. As the color sets became more popular the daily period of color-casting was increased and color programs were put on in the daylight hours as well as at night. By the end of the 1960's, all prime-time shows on the major networks were in color. Additionally, color television started in a number of other countries. International live telecasting became a reality with two-way television by satellites between the United States and both Europe and Asia. A small hand-held color camera was developed for the space program and many color pictures were received from the moon.

A number of new radio astronomy observatories were set up during this period and larger and larger radio telescopes were built or under construction. The largest of these was a 1,000-foot diameter reflector with an area of over 18 acres. During this period radar echoes were received from all the planets as far as Mars. At some of these observatories, projects were organized in an effort to detect signals from intelligent beings anywhere in the Universe. The results of this effort were negative but a new type of star known as a pulsar was discovered. A number of these stars were discovered before the first visible one was detected. The pulsar emits pulses of very high intensity radiation at regular intervals.

The radio amateur made several major achievements in addition to the distance records. The most outstanding of these were the launching of an amateur-designed and -constructed satellite known as Oscar 1, the first amateur color television transmission, and the first amateur contact by moon bounce.

By 1960, the third-generation computers were coming out, led by Univac. The speed of the new Univac was nearly ten times as fast as previous models. Univac remained one of the leaders in the field for some years, until integrated circuit models came out with their great size reductions and reduced power consumption. More and more applications for computers were being proposed and applied. One of the earliest was for assisting in aircraft traffic control and providing almost instant information on aircraft reservations from a month to six weeks in advance.

Because of the tremendous speed of the computer, the circuits were idle for relatively long periods of time. For economical reasons it was advantageous to put these idle circuits to work. A number of companies formed computer nets where many individual users could feed a computer system on a time-shared basis and not require a computer of their own. Computer programming still required appreciable time and skilled programmers were in demand. An approach to the problem was to train more operators. To do this several simplified programming systems were developed. COBOL and BASIC computer languages were developed during this period. Memory development continued with the aim of faster access and increased storage capability. By the close of the decade the holographic disc memory had been developed with a high degree of immunity from disc scratches or dirt and with a tremendous increase in storage capacity.

Besides such applications as payroll calculations, budget and inventory control, the use of computers was finding increased application by hospitals and doctors for rapid medical diagnosis and heart analysis. The computer had even gone into space with controlled photographic transmission of space objects and reassembly of the picture on earth. By the end of the decade computers had taken over control of some of the landing operations of returning spacecraft.

A major development in electronic circuits was the integrated circuit. At the beginning of the decade Texas Instruments introduced some simple logic integrated circuits closely followed by Fairchild. By 1962 the IC's were being mass produced with more and more companies getting into the integrated circuit business with its favorite logic type. By 1965, RCA entered the rush to integrated circuits.

Audio amplifier and computer memories were reduced to solid state and before the end of the 1960's a television set had been made completely with solid-state circuits except for the picture tube.

The miniature circuits were soon applied with the quartz-stabilized watches coming on the market with the Beta 21 and leading to the multifunction, extremely reliable and accurate watches on the market today.

In physics, progress was continuing on bigger and better

atom smashers. Several types of accelerators were being developed in a number of institutions including the Brookhaven National Laboratory, Stanford University, MIT and Harvard. Energies developed by these accelerators varied from six to 33 BeV (billion electron volts).

At the University of Pittsburgh a magnification of 200,000 times was obtained in their electron microscope. Photographs were released showing the atomic structure of metal surfaces. By comparing the photographs taken before and after irradiation of the surface, changes as small as the movement of a single atom could be seen.

From 1960 on, the development of lasers progressed from the observation of laser radiation from a synthetic ruby crystal to high-power gas lasers with 4,000 watts CW output or more. In France a Q-switched laser was developed with a pulsed output of 10^9 watts. Practical applications for the lasers were found almost from their beginnings. Their application to light-beam communication was apparent almost from the start. The Bureau of Standards used them for extremely precise distance measurements. By the middle of the decade, the laser was being used for eye operations and bloodless surgery. During the latter part of the period, the laser's powers had been increased to the point of their being tested for military use. The possibility of the laser was clearly demonstrated by shooting down a drone aircraft with a carbon dioxide laser. Shortly after the first laser was demonstrated, laser action in several types of solid-state material was observed. Before the end of the decade, this development had led to the light-emitting diodes introduced in 1968.

Although the wave nature of light had been known for many years, the idea of the hologram was conceived to improve the resolution of the electron microscope. Although some progress had resulted, the idea was soon put aside for some years because no satisfactory coherent light source was available at the time. With the development of the laser, an intense light source of high coherency became available and the announcement of the development of good holograms again revived interest in the field. More and more uses for the holograms were developed.

In the power field the major effort during this period was

the development of atomic power sources. At the opening of
the decade four stations in the United States were in operation. By the end of 1961 six new atomic plants were in operation. Canada's first atomic power plant came on line during this period. In the United States, there were about 15
atomic power plants in operation by the end of 1963 and in
certain areas atomic power was becoming cost-competitive
with that from fossil fuel plants. Some small strontium-90
generators were being used for remote locations such as
weather stations and lighthouses.

A number of hydroelectric plants were under construction and several new dams began generating power. A tidal
power system was demonstrated in France. Power could be
obtained from either the incoming or receding tides. Power
in Canada was being transmitted from remote sites at three
quarters of a million volts.

1960

AMATEUR

680. 10,000-MHz Record. The record for DX on 10,000
      MHz was broken by W7JIP/7 on Mary's Peak, Oregon
      when he made two-way communication with W7LHL/7
      on Green Mountain, Washington--a distance of 256-
      1/2 miles.

681. Color Television. What is believed to have been the
      first amateur color TV transmission in the US, and
      possibly the world, was put on 420 MHz by Melvil
      H. Shadbolt, WØKYQ in Dakota City, Nebraska.

682. Moon Bounce on 1296 MHz. A new distance record for
      1296 MHz communication was made on July 17 and a
      historic achievement was accomplished by members
      of the Rhododendron Swamp VHF Society, W1BU of
      Medfield, Massachusetts and the Eimac Gang Radio
      Club, W6HB of San Carlos, California. On that day,
      signals from W6HB were heard at W1BU over a path
      of about 475,000 miles by way of the moon. A few
      minutes later W1BU was heard at the California end

of the path. Four days later, signal reports were exchanged between the two stations. On July 24, Sam Harris, W1FZJ of Medfield, contacted W6HB by moon bounce.

BROADCASTING

683. Stereo Broadcasting. Six systems of FM stereo were field tested by KDKA-FM in Pittsburgh. The system finally adopted was that proposed by the General Electric Co. and the Zenith Corporation. Broadcasting was later authorized to start on June 1, 1961.

COMPONENTS

684. NPN Transistors. NPN triple-diffused transistors appeared with breakdown voltages up to 120v and gain bandwidth products of 10 MHz.

685. MOSFET. The metal oxide semiconductor field-effect transistor (MOSFET) was developed this year at Bell Laboratories. The MOSFETs were particularly useful as components of large-scale integration devices.

686. Kleimack, Loar, Ross and Theurer. The epitaxial method of transistor manufacturing was developed by J.J. Kleimack, H.H. Loar, I.M. Ross and H.C. Theurer of the Bell Telephone Laboratories. This manufacturing procedure made high-frequency transistors with relatively high-power output possible.

687. J.W. Allen and P.E. Gibbons. The British Physicists J.W. Allen and P.E. Gibbons developed point-contact gallium phosphide light-emitting diodes (LEDs) of various colors.

688. Joseph Sola. The constant voltage transformer developed by Joseph Sola in 1937 had been criticized because, although it regulated the voltage output, it did not maintain the sinusoidal waveform in the output and consequently some energy variation occurred in the regulated output. By 1960 the Sola Electric Company had improved the device so as to provide a

constant voltage output with very little distortion of the sinusoidal input.

689. Ceramic Tubes. The Cermolox line of power tubes for transmitters was put on the market by RCA. These were rugged and reliable metal-to-ceramic types.

690. Tubes. The 7270 high-perveance beam power tube was announced by RCA. It was rated at 315 watts input to 60 MHz or 235 watts cw to 175 MHz. It was unusually small for its power rating.

COMPUTERS

691. Univac III. Univac III, said to be nine times faster than Univac II, was introduced this year. It was the first computer to use the thin-film memory for its control memory.

INDUSTRY

692. Sony Corporation of America. The Sony Corporation of America went into business on March 1 as a sales organization for the Sony Corporation of Japan. After problems arose with their American distributors, the new Corporation was set up in this country by Akio Morita and Shokichi Suzuki.

693. Hewlett-Packard Co. In 1960 the Hewlett-Packard Company continued to expand by the acquisition of the Sanborn Co. of Waltham, Massachusetts and the F & M Scientific Corp. of Avondale, Pennsylvania. These acquisitions allowed Hewlett-Packard to apply electronics to problems within the medical and chemical fields. An additional plant was set up also in Loveland, Colorado. By the following year the company was getting into the field of solid-state components.

694. Teledyne, Inc. The Teledyne Corporation was founded by Henry Singleton and George Kozmetsky. Within ten years the company's electronic business exceeded $1 billion annually.

INSTRUMENTATION

695. Automatic IC Testing. With the increase in demand
     for more and more complex integrated circuit com-
     ponents, testing the devices became an increasingly
     costly process. Texas Instruments of Dallas put
     the first IC tester on the market. This instrument
     would make up to 36 tests on components with up to
     14 pins. Other companies came out with their test
     instruments. Teradyne brought out a diode tester
     and an automatic tester for resistors. Fairchild
     brought out the first of a series of transistor test-
     ers.

INTEGRATED CIRCUITS

696. Texas Instruments, Inc. In March, Texas Instruments
     announced the first custom-integrated circuit built
     for military equipment.

LASERS

697. Schawlow and Townes. The possibility of developing
     light amplification had been considered ever since
     the development of the maser. The general physical
     requirements and conditions for laser emission were
     studied this year in the US by A.L. Schawlow and
     C.H. Townes and showed the feasibility of several
     laser systems.

698. Theodore H. Maiman. Dr. Theodore H. Maiman of the
     Hughes Research Laboratories in July produced the
     optical laser emission from a synthetic ruby crystal.
     Work on the laser was also being done at Bell Labs,
     American Optical Co., RCA Laboratories, General
     Electric and Varian Associates, as well as by the
     air force.

699. Javan, Bennett and Herriott. The development of a
     successful helium-neon continuous wave laser was
     announced by A. Javan, W.R. Bennett, and D.R.
     Herriott of the Bell Telephone Laboratories.

PHYSICS

700. Metallic Glass--Pol Duwez. Metallic glass was discovered about this time by Professor Pol Duwez and associates at the California Institute of Technology in Pasadena. They discovered that if certain metallic alloy liquids were cooled fast enough, roughly at 1 million degrees per second, they would not solidify in a crystalline form. The metallic glass promised to provide a means of obtaining higher efficiencies in motors and transformers.

701. Atom Smasher. The particle accelerator or atom smasher of the Brookhaven National Laboratory was completed on Long Island in 1960. The alternating Gradient Synchrotron, or AGS as it became known, developed beams having energy levels up to 33 billion electron volts (33 BeV). This was the most powerful unit of its kind up to this time.

POWER

702. Commonwealth Edison Co. The first nuclear power plant financed privately by Commonwealth Edison was opened near Dresden, Illinois. This was a boiling-water reactor designed by General Electric. By the end of 1962, it had generated over 2 billion kwhr (kilowatt-hours) of electricity.

703. Nuclear Power. By the end of the year, there were four commercial atomic power plants in operation in the United States.

RADAR

704. Radar Warning Receivers. The first use of radar warning receivers in US aircraft occurred this year. The real value of such receivers in military operations was demonstrated in the Vietnam War. By 1971 virtually all US tactical aircraft were equipped with these detectors.

RADIO ASTRONOMY

705. <u>Extraterrestrial Intelligence</u>. An attempt was made at the National Radio Astronomy Observatory, Green Bank, West Virginia to detect the presence of life on other planets by scanning the heavens for radio signals indicating intelligence. This was done by monitoring many frequency channels with the antennas directed at nearby stars.

SATELLITES

706. <u>John R. Pierce</u>. The Echo experiment was proposed by Dr. John R. Pierce of the Bell Telephone Laboratories on hearing of the plan of NASA to launch a large balloon into orbit in order to study problems of orbital stability. He had previously published his calculations on the possibility of satellites for global communications. In August, the self-inflating satellite was launched into orbit. <u>Echo 1</u> was a 100-foot diameter, 130-pound aluminized plastic globe. On August 13 the balloon was put in a 1,000-mile orbit by NASA. Three days later the first transcontinental telephone communication was accomplished between Goldstone, California and Holmdel, New Jersey. The test confirmed the predicted path loss, stability, and performance but the balloon was soon destroyed by meteorites.

707. <u>Tiros Satellites</u>. Among the first satellites put in orbit for meteorological purposes were <u>Tiros 1</u> and <u>2</u>, launched on April 1 and November 23. <u>Tiros 1</u> took about 23,000 photographs. The camera of <u>Tiros 2</u> was defocused by the launch, but provided good infrared information. These were the first of a number of weather satellites launched by NASA.

TELEPHONE

708. <u>Electronic Switching (Telephone)</u>. In November, an experimental electronic telephone switching system was put into operation in Morris, Illinois. The system was not entirely satisfactory, but served to

demonstrate areas in which improvements were
needed.

709- TASI-A System. A time-assignment speech interpola-
10.   tion system was installed on the first transatlantic
      telephone cable by Bell Labs. This system, known
      as TASI-A, almost doubled the capacity of the line.
      During the natural pauses in the normal speech, the
      line would switch to the other party. The operation
      was so fast, it was not detectable to either party.

711. Transatlantic Telephone. It was reported that approx-
      imately three million transatlantic telephone conver-
      sations were completed this year.

      TELEVISION

712. Color Camera Tubes. On May 6, the model 4001 color
      tube was announced by RCA. The tube required no
      more light than a black-and-white picture tube. In
      December, a new color tube producing a picture
      about 50 percent brighter than previous tubes was
      introduced.

713. French Standards. France changed its standard TV
      format from 819 lines to 625 lines as was more gen-
      erally used in Europe and the United Kingdom.

      TIME

714. WWVL. The station WWVL began sending time signals
      in April. This is a VLF station operating on 20
      kilocycles (KC) and located at Sunset Canyon, Colo-
      rado. The low-frequency signal eliminated the errors
      accompanying ionospheric propagation.

715. Electronic Watches. The Bulova Watch Co. was the
      first to utilize microminiature electronic techniques.
      It brought out its accutron tuningfork watch in Oc-
      tober.

TROPOSPHERIC SCATTER

715a. Tropospheric-Scatter Communication. A tropospheric-scatter communication system from Florida to the Bahamas was put in service this year. The system provided 24 voice circuits. Other systems were put in operation in Algeria and Hawaii.

1961

AMATEUR

716. Oscar 1. Radio amateurs entered the space age with the first orbiting satellite carrying amateur radio (Oscar-1). This was the first noncommercial, non-governmental satellite, and was designed and built by amateur radio operators. Oscar 1 was launched December 12 by hitchhiking on the Thor-Agena rocket carrying the Discoverer 36 satellite. Oscar 1 achieved a 91-minute orbit and operated 23 days on the internal batteries. The launch date, December 12, was just 60 years after Marconi received his first transatlantic signals. Weight at launch was about ten pounds.

APPLIANCES

717. Automatic Laundry Dryer. Automatically controlled clothes dryers were developed this year. By monitoring the moisture in the fabrics, the machine would automatically shut off when the proper degree of dryness had been reached.

718. Appliances, Cordless. G.E. and Black and Decker, taking advantage of the storage capabilities of the new batteries, brought out the cordless electric toothbrush and the cordless electric drill.

BROADCASTING

719. <u>Nathan B. Stubblefield</u>. On May 18, the Kentucky Broadcasters Association presented a plaque officially recognizing Nathan B. Stubblefield of Murray, Kentucky as the inventor of Radio Broadcasting. The presentation was made to Mr. James L. Johnson, Executive Secretary of the Chamber of Commerce of Murray as custodian for the people of Calloway County. The plaque was on display in the Chamber of Commerce building in Murray.

720. <u>FM Broadcasting</u>. On April 19, the FCC announced the approval of FM Stereo Multiplexing, bringing a new interest in FM broadcasting. Approval became effective June 1, specifying the system recommended by General Electric and Zenith. The use of two stations and receivers for stereo would no longer be required.

721. <u>Stereo Broadcasting</u>. A National Association of Broadcasters survey indicated that an estimated 79 stations would be presenting stereo programs by the end of the year and over 175 by the end of 1962.

COMMUNICATIONS

722. <u>Canadian Microwave System</u>. A microwave communication link was put in operation between Grand Prairie, Alberta and Mount Dave on the Yukon-Alaska border. This was considered to have been the world's most advanced microwave system, consisting of 42 relay stations at intervals of five to 40 miles and providing both telegraph and voice channels.

723. <u>Satellite Communication</u>. NASA and AT&T signed an agreement to put an experimental communication satellite in orbit in 1962. Tracking would be done by the station at Andover, Maine. The satellite would relay data, telephone, and television from across the Atlantic.

COMPONENTS

724. Unipolar Field Effect Transistor. The unipolar field-
     effect transistor was put on the market by Texas
     Instruments. The current control in the transistor
     was modulated by the application of an electric field
     and consequently its operation was much like a
     vacuum tube. Input impedance is in the megohm
     region for the lower frequencies.

725. Compactrons. The General Electric Company announced
     its new development of tube technology in the form
     of the so-called compactron. Several prototype mod-
     els were exhibited. The tube required a 12-pin
     socket and one compactron could perform up to five
     tube functions. The greater space efficiency and
     lower cost, considering the number of tubes they
     replaced, made them particularly attractive to TV
     manufacturers.

726. Super Power Tubes. Super power grid-controlled
     amplifier tubes came out this year. The 7835 tube
     had 150,000 watts average plate dissipation with
     peak pulse power at 250 MHz.

727. Color Tubes. The three-gun color tube, utilizing
     shadow mask construction in a rectangular tube, was
     announced by Motorola. The deflection angle was
     92° for the 23-inch tube, permitting a considerable
     reduction in set depth.

728. Nuvistor Tubes. The millionth nuvistor tube was pro-
     duced this year. The tubes were used in some mili-
     tary and industrial equipment. The nuvistor was a
     very small tube developed to compete with the tran-
     sistor. These included the 6CW4, 7586 and 7587
     types.

729. Tubes. Eimac announced the first high-power zero-
     bias triodes. The 3-400Z with 400 w dissipation,
     the 3-1000Z with 1000 w dissipation and the 3CX10,
     000A7 with 10,000 w dissipation.
          Sylvania announced the improved version of the
     6146. This was the 6146A. The new tube handled
     the same power as the 6146 but with as low as five
     volts on the heater.

ELECTROSTATICS

730. A.N. Gubkin. The first electret electrostatic motor
was built by A.N. Gubkin. The electret motors
operated on the principle that like charges repel
and unlike charges attract. The electret is a "fro-
zen" electric field and produces an electrostatic field
just as a magnet produces a magnetic field.

INSTRUMENTATION

731. Function Generator. Wavetek introduced its Model 101
function generator this year. It provided square,
ramp, sine wave and triangular waveforms at frequencies less than 1 CPS to 10 MHz.

732. Electron Microscope. Magnifications of from 300 to
200,000 times have been obtained from the electron
microscope at the University of Pittsburgh. This is
believed to have been the most powerful microscope
up to this time.

INTEGRATED CIRCUITS

733. Integrated Circuits. Integrated circuits (ICs) were
put on the market this year by Texas Instruments
with its circuit series 51. These were simple logic
circuits. By the end of the year, integrated circuits in production quantities were being produced
by both Texas Instruments and Fairchild.

LASERS

734. Laser Communication. An experimental light beam
communication system at the Bell Laboratories was
successful over a 25-mile path.

735. Laser Development. Laser action from solid-state material was observed at Bell Laboratories from a neo-
dymium-doped calcium tungstate crystal. Announcement was also made of the first laser operation with
a helium-neon gas mixture. By March of this year,

IBM developed the samarium-doped and uranium-doped calcium fluoride lasers.

PATENTS

736. <u>Robert Noyce</u>. On April 25, the patent for a Semiconductor Device and Lead Structure was issued to Robert Noyce and described vacuum-deposited strips for connecting semiconductor circuits. The patent was contested by Texas Instruments. But the claim was upheld by the US Court of Customs and Patent Appeals (November 1969). This was a step in the development of large-scale integrated circuits.

POWER

737. <u>Statistics</u>. By the end of the year, eight commercial nuclear power plants were supplying commercial power in the United States.

738. <u>Atomic Power</u>. In August, the first operational strontium-90 fueled thermoelectric generator, developed by Martin Marietta, was installed on Axel Heiberg Island (north of the Arctic Circle). This was a 5-watt generator which provided power for an automatic weather station. It operated until 1965.

739. <u>DC Transmission</u>. DC cables were installed from Dungeness, England to Boulogne, France. The cables were energized with 200 kv DC this year. The lines were capable of supplying 160 kw of power.

RADIO ASTRONOMY

740. <u>Parkes Radiotelescope</u>. The 210-foot diameter radiotelescope, the largest in the southern hemisphere, was put in operation in Parkes, Australia. It was used for more detailed mapping of the arms of the Milky Way, the Andromeda Nebula and other celestial objects.

741. <u>Stanford University</u>. Before the end of the year, the

largest radiotelescope in the United States (150-foot diameter) was put in operation at Stanford University.

742. Mercury Radiation. Natural radio noise was detected coming from the planet Mercury at the University of Michigan. At that time noise had also been detected from Jupiter and Saturn.

743. Radio Astronomy. In January, scientists at the California Institute of Technology located the first visible radio star other than the sun. It was in the constellation Triangulum identified as 3C-48. This was the first time astronomers had been able to both hear and see the star emitting the signals.

744. Venus Detected. Microwave signals were bounced off of Venus in March by the Gladstone Tracking Station in the Mojave Desert in California. This had been done the year before but could not be verified at that time.

SATELLITES

745. Echo 1. Analysis of the data received from Echo 1 before its destruction showed that the signals could be reflected to any point in view of the satellite. The satellite data gave a more precise measure of atmospheric density and the influence of solar radiation on the atmosphere.

SEMICONDUCTORS

746. Wentorf and Bovenkerk. Dr. Robert H. Wentorf, Jr. and Harold P. Bovenkerk of General Electric's Metallurgical Products department were the first to produce semiconducting diamonds.

SPACE

747. SNAP Generator. The first space application of the SNAP Generator, a portable nuclear power source,

was used on the satellite <u>Transit IV-A</u>, launched
June 29.

TELEVISION

748. <u>Video Tape Expander</u>. Engineers of the American
Broadcasting Company developed the "VTX" which
permitted instant replay of any portion of a recorded TV pickup.

TIME

749. <u>Time Synchronization</u>. Synchronization of the time
signals of the United Kingdom and the United States
started January 1, 1961. Corrections of 50 or 100
ms were made simultaneously as necessary.

TRAFFIC CONTROL

750. <u>Traffic Control</u>. The Chicago area expressway surveillance project was set up to study the problems
of freeway traffic control. Electronic methods were
developed to detect cars and control traffic.

TROPOSPHERIC SCATTER

751. <u>Tropo Systems</u>. Tropo systems were set up in Belgium,
Sweden and Japan.

<u>1962</u>

ACTS

752. <u>Communication Satellite Act</u>. The United States Congress passed the Communication Satellite Act. The
act declared the policy of the United States "to es-
tablish in conjunction and cooperation with other
countries as expeditiously as possible a commercial

communication satellite system." The agreements were signed by 11 nations on August 20, 1964. The act created the Communications Satellite Corporation (COMSAT).

AMATEUR

753. <u>Oscar 2</u>. The second Orbiting Satellite Carrying Amateur Radio (<u>Oscar 2</u>) was launched June 2 into a 91-minute orbit. It was similar to <u>Oscar 1</u> and operated 18 days.

754. <u>Moon Bounce</u>. Two-way moon-bounce communication was established between amateurs on the east coast of the United States and Hawaii. This was accomplished on August 9 when W1BU contacted KH6UK on 1296 MHz.

BATTERIES

755. <u>Microbial Battery</u>. In 1962 a biological battery or fuel cell was reported to have produced about 900 milli-volts of electricity. This research was being conducted with the idea of developing power from waste, garbage or sewage, to provide power for space craft or isolated communities. Sufficient power has been developed to run a light bulb, radio, or a model-size motor.

COMPONENTS

756. <u>J.E. West and G.M. Sessler</u>. The foil-electret microphone widely used in "lapel" microphones, hearing aids and other small microphone requirements was invented by J.E. West and G.M. Sessler of Bell Laboratories. The microphones provided high sensitivity, fidelity, and reliability at low cost.

757. <u>JFET Transistors</u>. High-voltage junction field-effect transistors with breakdown voltages of over 300 volts came out this year.

758. <u>Metal Oxide Semiconductor Transistors</u>. Fairchild produced its metal oxide semiconductor (MOS) transistor
this year.

759. <u>Thyristors and Silicon-Controlled Rectifiers</u>. Thyristors and Silicon-Controlled Rectifiers (SCR's) appeared on the market as a consumer product this
year.

760. <u>Photo Cells</u>. Philco developed and demonstrated a
silicon planar epitaxial photodiode with a response
to 5 GHz.

761. <u>Photo Cells</u>. Camera manufacturers were turning to
cadmium sulphide photo cells for their exposure
meters because of their great sensitivity at low light
levels. This application was considered the most
significant improvement in exposure meters in over
20 years.

762. <u>Tubes</u>. Among the new tubes out this year were:
Amperex--8032, 8117, 8300; Eimac--4CX3000A,
4KM70SJ (this is a 20kw Klystron tuneable from 1700
to 2400 MHz); Mullard Master 10 Series (guaranteed
10,000-hour life); and RCA--8032, 8070, 8072, 8121,
8122, 6GJ5.

COMPUTERS

763. <u>Speech Study</u>. The Air Force Cambridge Research
Center was using a computer in the speech research
laboratory to break down the human voice spectrum
into its basic patterns. Voice had been broken
down into 400 patterns at that time. The patterns
may have simplified the generation of synthetic
speech.

764. <u>Computer Applications</u>. By 1962, airlines were using
computers for air traffic control and seat assignments up to a year in advance. A computer system
was coordinating reservations between 40 cities in
the United States.

765. <u>Benjamin Slatin</u>. The computer was adopted by the

paper manufacturers, according to Benjamin Slatin
of the American Paper and Pulp Association. The
computer was used to monitor several variables of
the paper-making process and type out required information on a remote readout at regular intervals.

INDUSTRY

766. Industry. In March, the Bell System founded Bell-
comm Inc. The company was jointly owned by the
American Telephone and Telegraph Company and the
Western Electric Company. The company was founded
because of a request by NASA for assistance in the
United States space exploration program. The company's first efforts were in performing space system
engineering studies for the space missions.

767. Holmdel Laboratory. The Holmdel Laboratory in Holmdel, New Jersey was opened by the Bell System primarily for radio research. It soon grew to become
the largest of the Bell Laboratories.

INFORMATION STORAGE

768. Stock Quotation. The Ultronic Systems Corporation
demonstrated a system for providing abstract instantaneous information on stock market prices or
roughly 40,000 other facts on 4,000 stocks. Information is sent by wire to subscribers.

INSTRUMENTATION

769. Cable Fault Locater. A British-designed fault locater
for submarine or underground cables had been announced. The echo caused by the fault was detected
much as in radar and was indicated in nautical miles.

770. Warren Macek and D.T. Davis. A laser gyroscope
was developed at the Sperry Gyroscope Co. in 1962
by Warren Macek and D.T. Davis. Some years were
required to reduce errors and increase its reliability.

771. **J. Bonanomi.** A thallium atomic-beam time standard
was built under the direction of J. Bonanomi at
Neuchâtel Observatory in Switzerland.

INTEGRATED CIRCUITS

772. **Integrated Circuits.** The mass production of integrated
circuits started this year. Transistor-Transistor
Logic (TTL) circuits were developed by Pacific Semiconductors Inc. Diode Transistor Logic (DTL) circuits were pushed by Signetics. Motorola developed
emitter coupled integrated circuits and introduced
the MECL-1. Texas Instruments brought out the
5400 series of IC using Transistor-Transistor Logic.

LASERS

773. **R.N. Hall and M.I. Nathan.** R.N. Hall and his assis-
tants of the G.E. Research Laboratories made the
semiconductor laser (September). Ten days later,
a team under M.I. Nathan at the IBM Research Lab
also made a semiconductor laser. These devices led
to the development of the light-emitting diodes introduced on the market in 1968. Researchers at the
Bell Laboratories and the Lincoln Laboratory of MIT
also reported observing gallium arsenide light emission.

774. **Lunar Laser Reflection.** The reception of bursts of
light from a laser which were reflected back to earth
from dark portions of the moon was accomplished
this year.

775. **Laser.** At the Bureau of Standards Research Laboratory, a helium-neon laser was constructed which
could theoretically measure a distance up to 60 miles
with a precision of one part in a million.

ORGANIZATIONS

776. **Institute of Electrical and Electronic Engineers.** The
Institute of Radio Engineers (IRE) merged with the

American Institute of Electrical Engineers (AIEE) and became the IEEE--The Institute of Electrical and Electronic Engineers.

PHYSICS

777. Brian Josephson. The Josephson junction, discovered by Brian Josephson, could provide switching speeds of ten picoseconds with power dissipation down in the microwatts. Nearly two decades later, it was being considered for computer application.

778. Field-Ion Microscope. Direct visual observation of the atomic structure of metal surfaces capable of showing the change after irradiation of a single atom was made possible this year by the field-ion microscope.

779. Electron Accelerator. Stanford University was awarded the Atomic Energy Commission Project to build the largest linear electron accelerator up to this time. Initially it would be built for 10 to 20 BeV with the possibility of doubling the acceleration in the future.

780. Cambridge Electron Accelerator. The Cambridge Electron Accelerator (CEA) financed by the AEC and built by scientists of MIT and Harvard University was completed. Electron acceleration providing energies of six billion electron volts was reported.

POWER

781. Nuclear Power. The nuclear power plant at Indian Point, Buchanan, New York on the Hudson River, was put in operation by Consolidated Edison on August 2 with a 275,000 kw unit. The project was started in 1954.

782. Antarctic. Atomic power for the McMurdo Sound naval base was turned on July 10. The reactor provided 1,500 kw with refueling expected to be required after two to three years' operation. The power was used primarily for the desalinizing of sea water for use at the base.

783. <u>Nuclear Power</u>. The first nuclear power station in
     Canada was put in operation at Rolphton, Ontario.
     The plant could provide up to 20,000 kw.

784. <u>Statistics</u>. By the end of the year, commercial power
     was being supplied by ten nuclear plants in the
     United States with several others being run by the
     government.

785. <u>Transmission Line</u>. About the middle of the year, a
     345 kv transmission line from Niagara Falls to New
     York City was put in operation.

RADIO ASTRONOMY

786. <u>Atmospheric Study Radar</u>. The Bureau of Standards
     in cooperation with the Instituto Geofísico de Huan-
     cayo had put in a radar for atmospheric ionization
     studies at a site about 17 miles east of Lima, Peru.
     Designed for 50-MHz operation, the antenna con-
     sisted of 18,000 dipoles distributed over 22 acres.
     Half had N-S polarization and the rest E-W. Peak
     power of the transmitted pulses was six megawatts.
     This was known as the Jicamarca Observatory. Us-
     ing scatter radar techniques, electron densities were
     measured at altitudes as high as 4,000 miles. In
     December, radio echoes from the planet Venus were
     detected.

787. <u>Mercury Detected by Radar</u>. The first radar reflection
     from the planet Mercury was reported by scientists
     in the USSR.

788. <u>Green Bank, west Virginia</u>. What was the world's lar-
     gest moveable parabolic antenna was built by the
     National Science Foundation for the Radio Astronomy
     Laboratory at Green Bank. This was a 300-foot-
     diameter dish with a resolution about ten times that
     of the previous system. It swung only north and
     south along the meridian. A 147-foot-diameter para-
     bolic antenna, which would be completely steerable,
     was under construction at the same site.

REFRIGERATION

789. **Refrigeration.** Thermoelectric refrigerators were brought out by five firms this year. They did not become popular at this time because of the low efficiency of the thermoelectric converters.

SATELLITES

790. **Robert E. Sageman--Telstar 1.** On July 10, Telstar 1 was launched from Cape Canaveral, Florida at 3:35 AM by NASA. The requirement called for an orbit having a perigee of 500 miles and an apogee of 3,000 miles. The achieved orbit was 514 miles and a 3,040-mile apogee. With this satellite, the first transatlantic telecast was relayed to Europe. The satellite was financed and built by the Bell Telephone Laboratories. Telstar 1 weighed about 170 pounds and was powered by batteries which were charged by solar cells. It was put in orbit by a Thor-Delta rocket. Project engineer was Robert E. Sageman. A number of firsts were achieved by this satellite. It permitted both live and taped television from Europe with color TV tests, two-way telephone digital data transmission and facsimile. Operation was good for about four months before its failure (attributed to high-energy electrons in the Van Allen Belt). It resumed operation the following year for a seven-week period.

791. **Relay 1.** A satellite built by RCA and known as Relay 1 was launched December 13. Relay 1 was an experimental satellite and was used to relay television to Japan, Europe and South America. The relay was made reliable by redundancy. Almost any failure could be bypassed by a redundant circuit. It was successful in replaying TV programs from the United States to South America and Japan in 1963. It operated only 200 days and completed over 500 communication experiments.

792. **Orbiting Solar Observatory (OSO-1).** The first study of the sun free of atmospheric absorption was made by OSO-1, launched March 7 from Cape Canaveral,

Florida. Until November 1963, <u>OSO-1</u> provided useful information on X rays, gamma rays, micromete-
orites, neutrons, the Van Allen Belt and others.
<u>OSO-1</u> received solar cell damage after high altitude
nuclear tests on July 9, but provided intermittent
but useful data until November.

793. <u>Ariel</u>. <u>Ariel</u> was launched from Cape Canaveral April
26 by a <u>Thor-Delta</u> rocket. <u>Ariel</u> was built as a
cooperative venture between the United States and
Britain with NASA responsible for the spacecraft
and power supply while British scientists designed
and built the experimental equipment. The satellite
was intended for ionospheric studies and operated
until nuclear tests in July damaged the solar cells.
Little or no useful data were received from <u>Ariel</u>
after September.

STANDARDS

794. <u>Standards</u>. An advance in the electrical measurements
field was a stable 100 million-ohm resistor to permit
DC voltage measurements up to 100,000 volts to be
made to within 20 parts per million accuracy. A
one-picofarad, 350,000-volt air capacitor was built
at the Bureau of Standards permitting 30 parts per
million accuracy in the calibration of instrument
transformers used on 350,000-volt power lines.

TELEPHONE

795. <u>Pulse Code Modulation</u>. The first large-scale digital
transmission system known as T1/D1 was put into
operation by the Bell System. The system multi-
plexed the voice signals using pulse code modulation.
The PCM system became commercially feasible with
the development of high-speed stable transistors.
The system transmitted 24 conversations at a digital
rate of about 1.5 megabits per second. It could be
used up to a range of about 50 miles.

796. <u>New Systems</u>. The Bell System put several new services in operation for data and voice with the Wide

Area Telephone Service (WATS), Wide Area Data
Service (WADS), and a private-line multitype service
called Telepak.

797. <u>All Electronic Telephone</u>. An experimental all-electronic
telephone system was put in service in Etna, New
York. Connections were made almost instantaneous-
ly. If the line was busy, the system would remember
the caller's number and automatically call back when
the line became clear.

TELETYPE

798. <u>TWX Network</u>. In August, the Bell System teletype-
writer exchange network in the United States was
revised to provide dial operation. This change was
reported to have saved about half of the time previously required.

799. <u>All-Channel Law</u>. Congress ruled that after April 30,
1964 no new TV sets could be shipped in interstate
commerce (in essence, no TV set could be manufactured) unless it could receive UHF as well as the
usual VHF channels.

TIME

800. <u>J. Bonanomi</u>. A thallium atomic-beam time standard
was built by J. Bonanomi at the Neuchatel Observatory in Switzerland. A few months later a thallium
standard was put in service at the National Bureau
of Standards. It demonstrated that thallium provided
as good an accuracy as cesium for similar-sized instruments.

801. <u>Quartz-Stabilized Watch</u>. The first quartz-stabilized
watch using integrated circuits came on the market.
It was known as the Beta 21.

TRANSPORTATION

802. <u>Trolley Cars</u>. Trolley car service was taken over by

buses this year in Baltimore, Maryland. Street cars had served the city just over 88 years.

1963

ANTENNA

803. Satellite Horn Antenna. The largest horn antenna ever built up to this time was installed at the Satellite Tracking Station at Andover, Maine this year. It had a mouth area of 3,600 square feet. It was 177 feet long and weighed 380 tons. The horn was protected by an inflated fabric radome.

APPLIANCES

804. Silicon-Controlled Rectifiers. One of the first applications of the silicon-controlled rectifiers (SCRs) was in the Whirlpool clothes dryer. The SCR was used to provide a continuously variable rotation speed.

COMMITTEES

805. Coordinated Universal Time. A meeting of the International Radio Consultative Committee (CCIR) was held in Geneva, Switzerland to consider time and frequency standards. At its recommendations, the system of coordinated universal time (UTC) was adopted.

COMMUNICATION

806. Project West Ford. On October 21, 1961, the Air Force had attempted to disperse millions of 3/4-inch dipoles in orbit to study the feasibility and liabilities of such a reflective communication system. The first attempt proved unsuccessful due to failure of the dispersing mechanism. The second attempt was

made May 9, 1963 from Point Arguello, California, when an <u>Atlas-Agena</u> B rocket launched 90 pounds of copper dipoles. These were dispersed on command May 10 into a 3,700-km polar orbit. In about six weeks the dipoles had dispersed to a closed ring around the earth. Scatter communication was accomplished with voice and high-speed teletype. The lifetime of the dipoles was predicted to be less than three years. This effort was known as Project West Ford. Transcontinental orbital scatter communication with this system for voice, high-speed digital data transmission and teletype was demonstrated.

COMPONENTS

807. <u>Connectors</u>. A new connector was announced by the General Radio Company. The device had a number of valuable characteristics for measurement laboratories. The connector was designed for 50-ohm coaxial cables. The VSWR was better than 1.01 up to 9 GHz. All contact surfaces were silver or a silver alloy for low resistance.

808. <u>Transistors</u>. Epitaxial-base power transistors were developed this year with gain-bandwidth products of about 10 MHz. General Electric Co. developed the first plastic-encapsulated silicon planar transistors for commercial use. NPN high-voltage transistors were developed with a breakdown of 80 v, and a gain-bandwidth product of 2 MHz.

809. <u>J.B. Gunn</u>. The gallium arsenide Gunn diode was discovered at IBM by J.B. Gunn. The material was also used at IBM to develop the Schottky barrier transistor (FET) which was the first satisfactory transistor for use at microwave frequencies.

810. <u>Tubes</u>. Tubes introduced this year included: Amperex 8179, a graphite anode tetrode for a single sideband operation; Nuvistors (RCA)--2CW4, 2DS4, 6DS4, 13CW4, 7895, 8056, 8058; Sylvania--low-noise strapframe grid tubes 6GK5, 7963, and 8186 (these tubes had a lower noise figure and higher gm than conventional tubes and offered improved performance in

circuits at frequencies up to 400 MHz); and United
Electronics--zero-bias triode UE572A was designed
to replace the 811A in single sideband circuits, capable of handling twice the power.

COMPUTERS

811. Computer Applications. By 1955, electronic computers
were beginning to find more and more applications.
In 1955 an estimated 50 computers were built by several companies. By 1963 approximately 10,000 units
had been built. They were put in use by many
stores and manufacturing companies for payroll and
budget calculations, inventory control, customer
billing and other management functions. New applications were being discovered every day. By this
year, computers were coming into use for medical
diagnoses. Law firms, libraries, and nearly all
facets of endeavor were finding more ways of operating their business more efficiently. The first-
known use of the computer in a public election is
believed to have occurred in Omaha, Nebraska on
April 23. Cards marked by the voters were fed
into the computer and the results were indicated
shortly after the polls closed. The use of computers
by the military was not new at this time, but the
application for the detection of live missiles from
decoys was now a possibility. Decoys up to 40 or
50 miles away could now be detected and their launch
point could be determined by computer analysis of
the trajectories.

812. Mini Computer. The Digital Equipment Corp. produced
the first commercial mini computer, the PDP-5. It
had a 12-bit, 1 k memory. This device started the
trend toward the smaller, more specialized computers.

813. P.J. Van Heerden. In 1963, a paper by P.J. Van
Heerden ("A New Method of Storing and Retrieving
Information"--Applied Optics, Vol. 2, 1963) pointed
out that the holographic technique could be used for
information storage utilizing the depth of the emulsion permitting up to 300 times more information in a
given emulsion than possible with straight photography.

CONFERENCES

814. ITU Administrative Radio Conference. The Administrative Radio Conference of the ITU was called primarily because of the expanding space communication and research operations. Delegates from about 74 countries and organizations attended. This group set aside the bands between 1400 and 1427 MHz exclusively for the study of radiation due to hydrogen in the galaxy--about 17 other frequency bands were allotted to radio astronomy. Over 20 bands were allotted for space research service.

ELECTRONIC IGNITION

815. Electronic Ignition. The mileage obtained from breaker points on an automobile could be extended roughly ten times by the use of an electronic ignition system. These systems were first used by at least one manufacturer in 1963.

HOLOGRAPHY

816. Emmett Leith and Juris Upatnieks. Emmett Leith and Juris Upatnieks at the University of Michigan produced laser holograms and began a spectacular development in the science of holography. Leith was awarded the "Man of the Year" award by Industrial Research for his work in laser holography. By the end of the year, they were making holograms of solid objects.

INTEGRATED CIRCUITS

817. Sylvania. Beginning this year, Transistor-Transistor Logic (TTL) circuits were produced by Sylvania.

PHYSICS

818. S. Shapiro. S. Shapiro is credited as being the first person to observe the oscillating supercurrent pre-

dicted by the equations of D. Josephson. He mixed a microwave signal with the oscillation and developed the difference frequency.

819. Superconductivity. A study of the possibility of using superconducting DC machines was started at the International Research and Development Co. Ltd. (IRD) in England. It was soon determined that only field coil superconductivity could be considered at this time.

POWER

820. Nuclear Plants. In the United States, there were 15 stationary nuclear power plants in operation and ten under construction or in the design stage by the end of 1963.

821. Nuclear Power. The first lighthouse operating on nuclear power was the unmanned station by Gibson Island, Maryland. It was powered by a strontium-90 source producing 60 watts.

822. Glen Canyon Dam. The 710-foot Glen Canyon Dam, the second highest in the United States, was completed in 1963 and the fill started. Glen Canyon is on the Colorado River near Page, Arizona. It was scheduled to produce power in 1964 (108,000 kw).

823. Steam Turbine. Steam turbine generator units capable of developing 650 MW were developed this year.

RADAR

824. Transistorized Radar. What was claimed to have been 'he first transistorized marine radar equipment was developed by Decca in England and introduced this year. An average of 1,000 sets per year were sold for the next four years. This same year another British firm, AEI Electronics, introduced a transistorized radar with a range of 15 yards to nearly 50 miles for use on ships.

RADIO ASTRONOMY

825. Radar Astronomy. The first radar signal reflections from the planet Mars were detected this year.

826. Venus Detected. The high power and sensitivity of the Jicamarca Observatory in Peru, jointly operated by the National Bureau of Standards and the Instituto Geofísico de Huancayo, made possible their first radar detection of Venus in December.

827. Puerto Rico Dish. The radio telescope with the 1,000-foot dish in Puerto Rico was dedicated in October. The antenna platform, 500 feet above the dish, weighed 600 tons. The motion, due to winds, was about one eighth of an inch, giving unexpected stability. The bowl was scooped out of a natural recession in the mountain. The dish area was over 18 acres. A moving feed mechanism permitted a scan area of over 40 percent. Director of the facility was William E. Gordon, originator of the idea. The station was said to be capable of detecting signals originating up to 12 billion light years in distance.

828. Radiotelescope. A new antenna for the radiotelescope at the University of Texas provided nearly 69 db gain at its operating frequency of 10,000 to 150,000 MHz. It was used to assist NASA in locating a suitable landing site on the moon for spacecraft.

RECORDING

829. TV Recorder. The first inexpensive recorder for TV programs was demonstrated in December. The recorder was developed in Great Britain and used one-quarter-inch tape moving at 120 inches per second.

SATELLITES

830. Syncom 2. The second Syncom satellite made by Hughes Aircraft Co. was launched in July into a geostationary orbit over Brazil. It was the first

satellite to use spin stabilization for axis stability.
It could handle telephone, facsimile and telex between the United States, Africa and Europe. It
provided the first satellite link with Africa. Over
this link, President Kennedy talked with the Prime
Minister of Nigeria, finishing his conversation with
"What hath God wrought"--the first message sent
over the Washington Baltimore Telegraph line by
Morse in 1845. Syncom 1 had been launched in
February 1963, but failed during launch and operated only about 20 seconds during an orbit adjustment.

831. Transit V-B. The US Navy's Transit V-B navigational
satellite was launched on September 28. This satellite was completely powered by a nuclear isotope
generator called the SNAP-9A (System for Nuclear
Auxiliary Power). The generator could provide 25
watts continuously for an estimated life of five years.
SNAP-9A had the advantage that it could not be af-
fected by space radiation.

832. Telstar 1. Telstar 1 had a transistor failure in November 1962. A way to bypass the transistor was
found, and operation was resumed on January 3.
It operated about seven weeks more before final
failure occurred. Failure was attributed to the high-
energy electrons in the Van Allen Belt.

833. Telstar 2. Telstar 2 was a replacement for Telstar 1.
Its transistors were mounted in an evacuated chamber, and on May 7 it was put in orbit at about
6,560 miles, a height above the Van Allen Belt for
most of its travel. It operated for a little over two
months before failure. About a month later, it resumed operation and continued to operate for at
least seven years.

834. Vela Hotel Project. With the signing of the nuclear
test ban treaty in 1963, Project Vela was set up by
the Advanced Research Projects Agency (ARPA) to
develop instrumentation for detecting nuclear tests
in the atmosphere, underwater, or in outer space.
On October 16 the US Air Force launched an Atlas-
Agena rocket carrying two Vela satellites from Cape

Canaveral, Florida. Initially the orbits were highly elliptical but later adjusted to approximately circular orbits at 57,000 and 100,000 miles.

835. **Weather Satellite.** The weather satellite, Tiros 8, was launched from Cape Canaveral on December 21. The camera took a picture every 208 seconds, stored it and provided a readout in 20 seconds, which was transmitted to earth.

STANDARDS

836. **Frequency Standards.** The General Radio Company announced new frequency standard instrumentation which would continue to operate in spite of normal power failure. The frequency drift at the end of one year of operation was one part in 10^9 per month.

TELEPHONE

837. **US-England Telephone.** Another Bell System telephone cable to link Great Britain and the United States was put in service. It was capable of handling about 140 simultaneous two-way conversations. Repeaters were required every 20 miles.

838. **Touch-Tone Telephones.** Late in the year, the first touch-tone telephones, with pushbuttons on the hand unit instead of a dial, were introduced. Service for this type of telephone was provided by over 200 central offices. By the end of the year, the service was used by about 150,000 customers.

839. **Hot Line.** An emergency communication link was established between Washington and Moscow on August 30. This line was to prevent or reduce the risk of war from a misunderstanding or accident.

840. **Submarine Cables.** About 5,000 miles of submarine cables were placed in service this year. A cable connecting the Canal Zone, Florence and Jamaica was put in service. The third cable from the United States to Europe (between Tuckerton, New Jersey

and Widemouth Bay, Cornwall, England) was laid by
the cable ship CS Long Lines.

841. Dataphone. The expanding demand for data transmission by the computer and other services was reflected in the telephone companies. AT&T-Bell System is reported to have installed around 5000 Dataphone modems to transmit digital information over
the telephone lines. Western Union had about 3,000
telex stations also transmitting data over the telegraph lines.

TIME

842. URSI. The General Assembly of the International Scientific Radio Union (URSI) recommended adoption of
a cesium transition as the standard for the physical
measurement of time.

TRANSMISSION LINES

843. Canada. Hydroelectric power in Canada was generally
being developed remote from the larger cities. This
was possible because of the high-voltage techniques
in use at that time for transmission lines. Power
was carried up to 500 miles from the source by transmission lines operating at 735,000 volts.

TRANSPORTATION

844. Sylvania Electric Products Inc. A system was developed by Sylvania to permit identification of railway cars travelling at 60 mph. The scanner read
the identifying color strips on the side of the car,
converted them to digits and provided a printed
record of the direction of travel and serial numbers
of the cars.

TROPOSPHERIC SCATTER

845. Tropospheric Scatter. Transatlantic communication by

tropospheric scatter was demonstrated. The system was said to have military value because of its secrecy. The voice signal was relayed by stations in Labrador, Greenland, Iceland and Great Britain.

1964

COMMUNICATION

846. Western Union. About the third quarter of the year, Western Union put in service a transcontinental microwave network with 267 relay and terminal stations. This was the longest single microwave project up to that time. The system provided an additional 50 million telegraph channel-miles capability. With all channels operating at full capacity, the system could handle 2.4 million words per minute.

COMPONENTS

847. J. Lüscher, H.C. Voorrips, B. Zega. A bidirectional silicon-controlled medium power switch was developed at the Battelle Memorial Institute, Geneva, Switzerland by J. Lüscher, H.C. Voorrips, and B. Zega. The device was a five-section structure with a control electrode on either end to control the direction of conduction. The switch operated much as two P-N/P-N switches in parallel but opposite directions. The overall voltage drop was about 1.5 v at maximum rated current.

848. Tubes. RCA announced the 6146A/8298A. This tube was an updated version of the 6146A and 8298 but permitted about 33 percent more power input. Amperex brought out the 8505 beam-power tetrode for mobile use to 250 MHz. Also the 8408, 8118 and 8348 instant-heating tetrodes for mobile use using a Harp-Cathode were introduced. RCA announced 6883B, 8032A and 8552 for mobile use.

COMPUTERS

849. RCA Spectra-70. In December, RCA announced its
      new computers utilizing the newly developed inte-
      grated circuits. These were the Spectra-70 family.

850. Computer Control. The idea of using electronic com-
      puters for direct control of processes in plants and
      factories appears to have developed about this time.
      A single computer with time-shared control could
      provide the necessary outputs to control the re-
      quired function as determined by properly located
      transducers serving as the input.

851. John G. Kemeny and Thomas E. Kurtz. The computer
      language called BASIC (Beginners All-Purpose Sym-
      bolic Instruction Code) was developed by Professors
      John G. Kemeny and Thomas E. Kurtz of Dartmouth
      College. The project was to develop a time-sharing
      system for General Electric computers to make the
      computers accessible to the faculty and students of
      the college. Work on the project resulted in the
      BASIC language in the spring of this year.

852. Computer Design. The use of computer graphics in
      automobile design began this year with both Ford
      and General Motors applying the technique in their
      new car designs.

853. IBM 360. IBM announced its system 360 series of
      third-generation computers on April 7. These com-
      puters attempted to standardize their instruction
      codes and arithmetic modes. The logic used discrete
      transistors which were much smaller than those used
      previously. These third-generation machines used
      solid state entirely, with increased memory size of
      over 500,000 characters.

854. American Airlines. American Airlines put into service
      a computerized reservation system. This system,
      known as SABRE, could process reservation requests
      in seconds rather than the average 45 minutes pre-
      viously required. SABRE handled all the reserva-
      tions for American Airlines in 55 cities at this time.

HOLOGRAPHY

855. <u>R.L. Powell and K.A. Stetson</u>. Powell and Stetson at
the University of Michigan Radar and Optics Lab,
are credited as being the first to notice holographic
interferometry in December, but it was noted almost
simultaneously at other locations. The effect was a
set of black lines in the hologram proving to be a
drift in position of the subject during exposure of
their hologram. This effect was also found at TRW
in Redondo Beach, California, and also at the National Physical Laboratory in England.

INSTRUMENTATION

855a. <u>Cesium Beam Standard</u>. The first portable cesium-
beam frequency standard was introduced by the
Hewlett-Packard Company in 1964. The tube length
in this standard had been reduced in size to 16
inches. A few years later, the tube length had
been reduced to about six inches.

INTEGRATED CIRCUITS

856. <u>Integrated Circuits</u>. Fairchild anounced its 930 series
of Diode Transistor Logic integrated circuits. These
were popular because of greater noise immunity than
other types. Fairchild also brought out its 702 operational amplifier. These devices were a vital link
in the development of large-scale integrated devices.

857- <u>Texas Instruments</u>. Texas Instruments brought out
8. its 5400 series ICs using TTL.

LASERS

859. <u>Atmospheric Research</u>. The laser this year was applied to atmospheric research. A beam projected into the atmosphere could be used to give cloud
heights, dust concentration and other information
required for atmospheric studies.

860. <u>Laser Tracking</u>. Laser tracking of satellites started
     with the launching on October 10 of <u>Explorer 22</u>.
     This was the first satellite to carry a retroreflector.
     The laser tracking was started by the Smithsonian
     Institution Astrophysical Observatory at Cambridge,
     Massachusetts. A laser reflection was detected by
     G.L. Snyder of General Electric Co. on October 18.

MEDICAL

861. <u>Drs. H.C. Zweng, M. Flocks, and Others</u>. The ap-
     plication of the laser for eye operations, for retinal
     tears, or "tenting" was reported. The action was
     so fast and painless, the patient could return to
     normal activities immediately. The first use of this
     technique was attributed to Drs. H.C. Zweng, M.
     Flocks, and others at the Stanford University School
     of Medicine.

862. <u>Dr. L. Goldman and R.G. Wilson</u>. The use of the
     laser for bloodless surgery in cancer operations was
     applied with good success by Dr. L. Goldman and
     R.G. Wilson of the Cincinnati General Hospital.

863. <u>Hearing Aids</u>. Zenith produced the first hearing-aid
     integrated circuit. It utilized six transistors and
     was smaller than the head of a match.

864. <u>Drs. D.H. Howry and J.H. Holmes</u>. Doctors Howry
     and Holmes of the University of Colorado developed
     the somascope--a device using ultrasound at 1,000
     pps that could produce a two-dimensional picture on
     a cathode-ray tube serving in some ways as an X-
     ray photograph without the radiation danger.

ORGANIZATIONS

865. <u>International Telecommunications Satellite Organization</u>
     (Intelsat). Intelsat was organized in August
     for the development of a global commercial communi-
     cation system. Initially 11 countries signed the in-
     terim agreement. Operation and technical services
     were the responsibility of the Communications Satel-

lite Corporation (Comsat). By the end of the year, 19 nations had signed. The first <u>Intelsat</u> satellite was launched in April of 1965.

PACKAGING-MANUFACTURING

866. <u>Bryant Rogers</u>. The Dual In-line Package (DIP) was invented by Bryant Rogers of Fairchild Semiconductor Co. this year.

867. <u>Martin Lepselter</u>. This year, Martin Lepselter of Bell Telephone Labs developed the "beam lead"--a combination mechanical support, heat sink and electrical conductor for transistors.

PHYSICS

868. <u>Townes, Basov and Prochorov</u>. Charles Hard Townes of the United States and Nikolai G. Basov and Aleksandr M. Prochorov of the Soviet Union received the Nobel Prize for development of the Maser and Laser Principle of producing high-intensity radiation.

369. <u>Barry Goldwater</u>. Senator Barry Goldwater disclosed to Congress that electromagnetic pulses from nuclear explosions could cause catastrophic failure in missiles and other electronic equipment. This disclosure resulted in a study of how to harden these components against the effects of radiation.

POWER

870. <u>EBR-2</u>. The Experimental Breeder Reactor #2 (EBR-2) was demonstrating power generation exceeding 30 MW this year. By 1969 it had reached its design power level of over 62 MW.

RADAR

871. <u>Over-the-Horizon Radar</u>. On September 17, President L.B. Johnson disclosed the development of an over-

the-horizon radar system. With this system any
Soviet-launched missiles would be detected within
seconds of their firings, allowing the warning time
for the United States to be doubled.

872. **Multifunction Array Radar (MAR)**. On July 1, a multifunction array radar (MAR) was put in operation at
the White Sands Missile range in New Mexico. Operation was so fast it could apparently look in all
directions at once. It was expected to replace the
previous target-acquisition, target-discrimination and
target-tracking radars.

SATELLITES

873. **Echo 2 Satellite**. The second Echo satellite was
launched on January 25. This was a passive satellite which was inflated to an approximate sphere 135
feet in diameter and chemically hardened to maintain
the shape. Echo 2 was launched January 25 from
the Pacific Missile Range and permitted communication experiments as well as the atmospheric density
and air drag at 700 miles altitude.

874. **Ion Propulsion**. The use of an ion engine for spacecraft propulsion was applied by the Russians this
year. The thrust generated by these engines was
estimated to have been several hundred newtons.

875. **Orbiting Geophysical Observatory (OGO-1)**. OGO-1
was launched from Cape Kennedy September 5 by an
Atlas-Agena rocket. OGO was contained in a 6' x
6' x 3' box. With its solar panels extended, it was
59 feet long with a 20-foot wingspan. It was designed to study cosmic rays, protons and electrons
in the Van Allen Belt, solar wind, the ionosphere
and upper atmosphere, VLF propagation, etc. The
satellite never stabilized properly but the stabilization problem did not prevent most experiments from
being completed successfully.

876. **Relay 2**. The second Relay satellite, built by RCA
under sponsorship of NASA's Goddard Space Flight
Center, was launched from the Air Force Eastern

Test Range, Cape Canaveral, Florida on January 21. The satellite successfully relayed pictures of the Winter Olympics from Austria and also live TV from Japan to the United States. The orbit varied from 1,294 to 4,604 miles from the earth.

877. Syncom 3. In August, the Syncom 3 satellite was put in synchronous orbit over the Pacific Ocean. With the satellite, the Olympic Games being held in Japan were seen in the United States (October). The Syncom Satellites were set spinning at 2 RPS to keep the axis pointing in one direction; this required only one antenna for transmitting and receiving. Four antennas were used for telemetry and command purposes.

878. Vela Hotel. The second pair of Vela satellites were launched on July 17 and were reported to have been put in coplanar orbits at 65,000 miles altitude. The specific purpose of these satellites was not announced at the time but they carried radiation monitors to check for cheating on the nuclear test ban treaty.

SEISMOMETER

879. Project Vela Uniform. Under the Vela Uniform Program, the Air Force Cambridge Research Laboratories and Texas Instruments developed an ocean-bottom seismometer for detecting underwater blasts or earthquakes. The third-generation device was tested at 24,000 feet with a 30-day recording capacity. The system was described in the December 1965 Proceedings of the IEEE.

SPACE PROBE

880. Ranger 7. The first live TV from deep space was made by Ranger 7 on July 28. Ranger 7 sent from the moon the most detailed photographs ever taken of the surface. Ranger carried six cameras designed by RCA for this specific purpose. Over 4,300 photographs of the lunar surface near Mare Cognitum were received.

TAPE PLAYERS

881. Tape Players. Four-track stereo tape players for auto-
     mobiles began to come into popular use about this
     time. In four or five years, they became one of the
     most popular items in the electronics field.

TELEPHONE

882. United States-Japan Service. A submarine cable be-
     tween the United States and Japan was put in serv-
     ice June 18. It extended about 5,300 nautical miles
     from Hawaii to Japan by way of Midway, Wake and
     Guam islands. At Hawaii it connected with cables
     to Canada and Australia.

883. Coax Cable System. A new coaxial cable system pro-
     viding approximately 9,000 telephone circuits to the
     transcontinental telephone system was completed by
     the Bell System. The system was designed to be
     practically immune to floods, hurricanes, or nuclear
     attacks. The system utilized 4,000 miles of cables
     with repeater stations underground.

884. Picturephone. The Bell System Picturephone service
     was reported to have been set up in the cities of
     New York, Chicago and Washington, DC.

TIME STANDARDS

885. Standards. The International Committee on Weights
     and Measures changed the time standard based on
     the earth's orbit around the sun to one based on
     atomic time.
        At the 12th General Conference, the Committee
     temporarily adopted the transition of the cesium-133
     atom as the standard of time. A value of
     9,192,631,770 Hz was assigned to the cesium transi-
     tion selected. This gave a second to an accuracy
     of one or two parts in 10^{13} representing a variation
     of one second in 300,000 years.

<u>1965</u>

AMATEUR

886. <u>Oscar 3</u>. The <u>Oscar 3</u> satellite, built by amateur radio operators, was launched on March 9 and put in a 103-minute polar orbit. It carried a two-meter transponder and beacon transmitter and operated for 15 days from its internal batteries.

887. <u>Oscar 4</u>. The fourth Orbiting Satellite Carrying Amateur Radar (Oscar 4) was launched December 21 by a <u>Titan 3 C</u>-rocket from the Eastern Test Range. The orbit was to have an easterly drift of about 30° per day. Altitude was about 18,000 miles with an orbital period of about ten hours. <u>Oscar</u> was built by the Radio Club of TRW, Redondo Beach, California. Due to damage received during the launch phase, it operated only about two weeks before failure.

COMMUNICATION

887a. <u>DATEL</u>. RCA Communications announced a new service for overseas data transmission known as DATEL. This service provided speeds of up to 1,200 words per minute. The system operated from punched cards, paper, or magnetic tape.

COMPONENTS

888. <u>The Triac</u>. The Triac solid-state AC switch was introduced by General Electric Co. This development was the result of a navy contract calling for "the development and evaluation of a bilateral switching device capable of carrying 25a."

889. <u>Capacitors</u>. A new form of capacitor was introduced by the Sprague Electric Co. This was known as the Filmite K Polycarbonate Film Capacitor. The low dissipation factor, high dielectric constant material permitted a size reduction of at least ten times,

and an essentially zero temperature coefficient capacitor.

890. Johnson, DeLoach, Cohen and Read. It was demonstrated this year by R.L. Johnson, B.C. DeLoach, Jr., and B.G. Cohen that specially made silicon P-N junctions when pulsed would create microwave oscillations and that these diodes were similar to those described by W.T. Read in 1958. The output was restricted because of high thermal resistance. These diodes were called IMPATT diodes because they operated in what is known as the Impact Avalanche and Transit Time mode of operation. The name was proposed by W.T. Read in an article published in the Bell System Technical Journal (February 1965).

891- Tubes. The following tubes were introduced this
2. year: Amperex--SD568, 7378, 8163, 8457, 8458, 8462, 8505, 8579, 8637; RCA--4604; and Scientific Instrument Research--572B.

COMPUTERS

893. Computer Time Sharing. One of the first commercial computer time-sharing systems was started by General Electric Co. The system permitted one computer to be used by many people at one time, all connected to the machine by telephone line. Within a few years many companies were offering remote computer services.

894. Mars Photographs. The most spectacular use of computers during the year was from Mariner 4 in close-up pictures of Mars. Computer-controlled photographs were taken and transmitted to Earth in 64 shades of gray. On Earth, the pictures were decoded by another computer and given the proper tones to reconstruct the picture.

895. Digital Equipment Corp. The Digital Equipment Corporation of Maynard, Massachusetts, brought out the first mass-produced computer to sell for under $20,000. This was the PDP-8 Programmed Data Processor.

HOLOGRAPHY

896. **Holograms**. The holograms of Leith and Upatnieks in 1963 revived the interest in holography. With the laser source of coherent light now available, many organizations were investigating the new techniques. Laser photography was developed about this time. In this process a three-dimensional object was photographed using coherent light. The reflected light waves were captured on film. This is called the hologram. The hologram bears no visual resemblance to the original object. By using coherent light and the proper techniques, an image of the object could be restored and seen in three dimensions, much as looking at the object directly.

INDUSTRY

897. **Lasers**. In December, the laser was put to production work at the Buffalo, New York plant of the Western Electric Co.

898. **Integrated Circuits**. RCA entered the integrated circuit market with a line of 17 types of ICs for military, industrial and instrumentation systems. This had grown to 25 types by the middle of 1966.

899. **MECL III Logic**. The fastest logic to date was brought out by Motorola with its MECL III Logic. Delay was reduced to the order of a nanosecond with a 60 mw gate dissipation. However, because of packaging problems and excessive power dissipation, it did not become popular.

LASER

900. **C.K.N. Patel**. C.K.N. Patel of Bell Telephone Laboratories announced the molecular gas laser. He found gas lasers were more efficient if the gas had heavy molecules. His highest efficiency was obtained with carbon dioxide gas.

PHYSICS

901. **Magnetism**. The strongest magnetic field ever de-
     veloped by man was generated at the National Mag-
     netic Laboratory of MIT. Using a water-cooled mag-
     net, a field of over 250,000 gauss was obtained.
     This was about 100,000 gauss higher than had been
     recorded previously.

POWER

902. **Atomic Power**. At about this time, because of the in-
     creasing costs of gas, oil, and coal, atomic power
     plants were becoming economically competitive with
     conventional power sources.

903. **Power Transmission**. This year, lines from the Manic
     dams in Canada began operating at 750 kv. This
     was the highest voltage used up to that time in Can-
     ada.

904. **The Great Blackout**. The largest blackout up to this
     time occurred on November 9, causing an area of
     about 80,000 square miles to be completely without
     power. It is estimated that about 30 million people
     in the United States and Canada were affected.
     Most of New York State, Massachusetts, Connecticut,
     Rhode Island, and parts of Pennsylvania and New
     Jersey, as well as a large part of Ontario, were
     without power for up to six hours. The cause was
     traced to the failure of a relay at a plant in Ontario.

RADIO ASTRONOMY

905. **The Arecibo Site**. The 18.5-acre reflector carved out
     of the valley at Arecibo, Puerto Rico was completed
     this year. The installation was made by the Ad-
     vanced Research Projects Agency (ARPA). This
     was the world's largest antenna dish, with a diameter
     of 1,000 feet. This served as a radar and radio
     telescope antenna. It was said to have detected sig-
     nals from sources 10 to 12 billion light years away.
     In July, amateurs at the site used the dish for a

moon-bounce experiment. The test was 100 percent successful as a number of single sideband contacts were made by moon bounce to both the United States and Europe. Operation was on 432 MHz.

SATELLITES

906. Orbiting Geophysical Observatory (OGO-2). OGO-2 was launched from Point Arguello, California on October 14, by a Thor-Agena rocket but jet problems prevented stabilization in orbit. The satellite, tumbling and not keeping its solar cells properly oriented to the sun, resulted in a varying power source. However, experiments could be made when sufficient power was available.

907. Orbiting Solar Observatory. The second Orbiting Solar Observatory (OSO-2) was launched from Cape Kennedy on February 3 by a Thor-Delta rocket. OSO-2 was to perform studies of the influence of solar phenomena on interplanetary and the terrestrial environment. Studies include monitoring of X rays, gamma rays and cosmic rays and their effects on zodiacal light. Tests continued to the end of November.

908. Early Bird (Intelsat 1). The Early Bird synchronous communication satellite was launched by NASA from Cape Kennedy, Florida on April 6 and put in stationary orbit over the Atlantic Ocean. At that time it provided communication between Andover, Maine and four stations in Europe. Commercial operation was inaugurated June 28. Early Bird provided 240-voice, two-way circuits. Initially 12 countries had telephone links with the United States. By 1966, 188 countries could be reached by Early Bird. It was initially expected to operate for 18 months, but it operated until January of 1969. Intelsat 1 was built by Hughes Aircraft Co.

909. Soviet Satellites. The first admitted communication satellite to be launched by the Soviet Union was the Molniya 1, launched April 23. The orbit was highly elliptical and placed to spend most of its travel over the Soviet Union. A second Molniya satellite was placed in orbit October 14.

SPACE PROBES

910. **Pioneer 6.** <u>Pioneer 6</u> was launched on December 16
from Cape Kennedy, Florida. Its purpose was to
investigate interplanetary space and send information to Earth on magnetic fields, solar wind, ionization, etc. The satellite carried two travelling wave
tube transmitters. The first operated approximately
122,000 hours before the backup travelling wave
tube had to be switched on.

911. **Ranger 8.** The <u>Ranger 8</u> space probe, developed by
RCA and the Jet Propulsion Laboratory, was
launched from the Eastern Test Range on February
17 by an <u>Atlas-Agena</u> rocket. On February 20,
<u>Ranger</u> landed on the moon in the Sea of Tranquil-
ity. During its life, it transmitted 7,160 pictures
of the surface to Earth.

912. **Ranger 9.** The <u>Ranger 9</u> spacecraft was launched to
the moon March 21. It impacted on March 24 in the
crater Alphonsus. Before impact, 5,814 pictures
were taken from an elevation of 1,300 miles to about
4,000 feet.

TELEPHONE

913. **Electronic Switching.** The first in-service test of the
all-electronic switching central station was started
May 30 in Succasunna, New Jersey. This was the
beginning of a complete changeover of the Bell System from electromechanical to all-electronic switching.

914. **United States-France Line.** The fourth transatlantic
cable between New Jersey and France was put in
service. The cable provided an additional 128 voice
circuits.

TELEVISION

915. **Color TV.** It was about the middle of the year when
color television viewing was available both day and
night and the demand for color sets became high.

In spite of their high cost approximately 25 percent of the sets sold were color sets. By the end of the year about 45 percent of the night programs were in color.

916. Cable Television. In 1965 the FCC established the first rules governing cable television for systems receiving signals by microwave. The following year the FCC established regulations for all cable systems, even though they were not served by microwave.

TIME

917. M.F. Easterling. About this time a method of time synchronizing of the various stations of the Jet Propulsion Laboratory by moon bounce was developed by M.F. Easterling. Time synchronization to within ± 20 microseconds was claimed. The radar operation was on 7150 MHz.

TRAFFIC CONTROL

918. Traffic Control. By this year, automatic ramp traffic signals and instruction signs were put in operation in the Chicago area.

VIDEO RECORDERS

919. Video Recorders. Early in the year, video recorders were perfected for home use for recording and playback of TV programs.

1966

COMPONENTS

920. Coaxial Switches. The Polyphase Instrument Company introduced a new series of coaxial switches; a single-

pole, two-position; a single-pole, five-position and a two-pole, two-position switch. All were rated at 1 kw, dc through 100 MHz at 50-70 ohms.

921. High-Power Switches. A line of high-power rotary R.F. switches was announced by the James Miller Manufacturing Co., Inc. These were part of the 51000 series. Included in the line were one- and two-pole, two to six positions and four-pole, two to three positions. All were rated at 20 amperes at 10,000 to 15,000 volts dc to 30 MHz.

922. Gunn Diodes. The first commercially available Gunn diodes were marketed by the International Semiconductor Inc. this year.

923. Tubes. New tubes out this year include: Amperex--8458; Eimac--5-500A, 4CX1500A, 4CX1500B, 4CV1500B, 5CX1500A; and Machlett Labs--PL 8583/267.

924. Color Tubes. By the end of the year RCA was making color picture tubes of 14-, 18-, 20- and 23-inch diagonals. The round tubes were no longer made by RCA. Round tubes were rapidly being replaced by the rectangular types in the new television sets.

COMPUTERS

925. A.H. Bobeck. The magnetic bubble memories, providing solid-state high-density memory capabilities for computers, was invented by A.H. Bobeck and associates at the Bell Laboratories.

926. Univac. The Univac Model 9200 and 9300 computers, disclosed in June, used thin-film-plated wire memories. These were the first computers to use the new magnetic memory technique. The plated wire replaced the semiconductor memory used in earlier Univac models.

927. Cybernet. Following the lead of General Electric, IBM opened the Service Bureau Company providing a time sharing link of 125 System/360 computers called Cy-

bernet. Remote computer operation was catching on and the Control Data Corporation opened another computer net later in the year. Within several years, dozens of other companies were offering time-shared computer services.

928. <u>Hewlett-Packard Company</u>. The demand for small computers was increasing. The Hewlett-Packard Co. of Palo Alto, California was one of the first to join the group of small computer manufacturers, with the Model 2116A introduced this year.

HOLOGRAPHY

929. <u>Holography</u>. A new application of holography was developed this year. This was a method of stress analysis by comparing the hologram made before and after the stress was applied.

INSTRUMENTATION

930. <u>Magnetic Tape Recorder</u>. A new type of tape recorder using current rather than voltage pickup from the tape was announced by Hewlett-Packard in December. The new recorder had a frequency response from about 400 Hz to 1.5 MHz. Advantages claimed for the system included wide bandwidth and signal to noise ratio greater than 30 db at about one percent distortion level. Distortion of the even harmonies was less than 0.1 percent.

931. <u>Solid-State Oscilloscope</u>. The first solid-state oscilloscope appeared on the market with the introduction of the Hewlett-Packard Model 180A. This instrument covered the range of DC to 50 MHz with nearly instant turn-on and other features equal or superior to those of large conventional high-frequency oscilloscopes.

INTEGRATED CIRCUITS

932. <u>RCA</u>. This year, integrated circuit amplifiers were be-

ing used for the audio stage in some TV sets. The
use of ICs was increasing in other consumer products as well.

PHYSICS

933. Linear Accelerator. A two-mile linear accelerator at
Stanford University was put in operation and produced electron beams with energies exceeding 18
BeV. This was approximately three times the energy
output of any previous linear accelerators. Electrons
in this accelerator increased their mass approximately
40,000 times in the trip down the accelerator.

934. J.R. Powell, Gordon R. Danby. Two scientists from
the Brookhaven National Laboratories in New York--
James Powell and Gordon Danby--proposed superconductivity for Maglev systems. Their research in-
spired more research in the United States, Japan,
Canada, Germany and Britain.

POWER

935. Plutonium Plant. The first privately owned plutonium
production plant in this country was put in operation
in West Valley, New York by Nuclear Fuel Services,
Inc. The plant operated until 1972 when it was
closed because of uneconomical operation.

936. Tidal Power. In France, construction had started in
1961 on a plant to harness the ocean tides by Electricité de France. Power was produced with the
tides going either in or out. The plant was completed this year and proved that such plants are
feasible. This plant was located in the Rance Estu-
ary near Saint Malo.

SATELLITES

937. Intelsat 2. The second Intelsat satellite, Intelsat 2,
built by the Hughes Aircraft Co., was put in orbit
over the Pacific Ocean October 26, but did not ob-

tain the correct position and drifted around. It
provided short-time communications but was soon
replaced by a second Intelsat 2 put in orbit January
14, 1967.

SPACE EXPLORATION

938. Surveyor 1. Surveyor 1, built by Hughes Aircraft
Co. made a soft landing on the moon on June 2, 63
hours and 36 minutes after earth launch. Surveyor
was launched to determine the possibility of a suitable landing area for a manned landing. The landing was made in the "Ocean of Storms." Earth had
received 11,348 pictures when the transmission
stopped suddenly.

STANDARDS

939. Resistance Standard. A new form of precision resistance was developed at the Hewlett-Packard Laboratory. It had an adjustment capable of setting the
value to less than one part per million.

940. Superconductive Motor. The first superconductive DC
motor was built in England by the International Research and Development Co., Ltd. The motor was
operated on June 1. It developed 50 HP at 2,000
rpm. This success led to plans for a larger machine
which was started the following year.

TELEPHONE

941. Venezuela Cable. A submarine telephone cable link
between the United States and Venezuela was opened
on August 3 with station-to-station calling introduced.

TELEVISION

942. TV Statistics. UHF television was becoming more popular. By 1966 approximately 40 percent of the TV

sets in use could receive UHF telecasts. Of the
new stations to start telecasting this year, two-
thirds were UHF stations. By the year's end, approximately 200 applications for UHF were on file at
the FCC.

943. <u>CFCF Television</u>. Color TV started in Canada on
July 1, when the Marconi Station CFCF-TV in Montreal came on the air.

944. <u>Transistorized TV Sets</u>. Fairchild Semiconductors
built a color TV set which was transistorized except
for the high-voltage rectifier, VHF tuner and vertical scan output. The first completely tubeless TV
set, except for the picture tube, is believed to have
been one developed by RCA this year. It was not
marketed at this time however.

945. <u>Color TV</u>. By the end of the year, all nighttime television programs broadcast by the National Broadcasting Company were in color except for an occa-
sional feature film in black and white.

TIME

946. <u>Time</u>. WWV in Greenbelt, Maryland went off the air
at OOOOGMT on July 1 and the new WWV in Fort
Collins, Colorado began operation. The new station
ran the transmitter at half-power for increased reliability. The station radiated 10 kw on 5, 10 and
15 MHz. Four 5-kw transmitters radiated 2.5 kw on
2.5, 20 and 25 MHz.

TRAFFIC CONTROL

947. <u>Traffic Control</u>. In the Los Angeles area, computer-
controlled on-ramp traffic control systems were being
installed at some freeway locations.

TRANSPORTATION

948. <u>Electrovan</u>. The Electrovan car built by General Motors

was powered by 32 fuel cells using hydrogen and oxygen in liquid form. It provided a 150-mile trip. The fuels hydrogen and oxygen, however, were too dangerous for practical use.

TROPOSPHERIC SCATTER

949. Tropospheric Scatter. A tropospheric-scatter system was inaugurated between Faial and Flores in the Azores.

1967

COMMUNICATION

950. Aircon. The Aircon System was introduced by RCA in 1967. This was a system permitting companies to plug into a master computer for automatic message relay to other associated organizations.

COMPONENTS

951. Color Tubes. The brightness of color tubes was greatly improved by new phosphors used in the RCA color tubes. Still greater improvements were made several years later.

952. Tubes. In transmitting tubes, the highest powered tetrode to date was brought out by Eimac, the 4CV250,000C and a smaller 100-kw version, the 4CV100,000C. Amperex brought out the 5894B/8737 and Cetron, the 572B/T-160L.

953. L. Armstrong. The IMPATT diode was improved this year by L. Armstrong using a zinc diffused, P-N junction mesa. With this diode, 85 mw CW could be generated at 10 GHz at about 10 percent efficiency. This was a GaAs diode.

COMPUTER

954. Integrated Circuit Computers. The PDP-8 computer, made by the Digital Equipment Corporation of Maynard, Massachusetts and brought out in 1965, was updated by a new integrated circuit model brought out this year.

955. Cecil H. Coker, Osamu Fujimura. By studying the movements of the mouth and vocal tract, Cecil Coker of the Bell Telephone Laboratories, with Osamu Fujimura of Tokyo University as a consultant, designed a computer-controlled electronic speech synthesizer producing very acceptable speech.

HOLOGRAPHY

956. High-Speed Holograms. High-speed pictorial holograms became possible with new film by Agfa-Gevaert Scientia. The emulsion had its sensitivity adjusted to match the radiation of the ruby, helium-neon or argon lasers.

INTEGRATED CIRCUITS

957. Very Large Scale Integration. It was about this time that interest was turning to the development of connecting even more circuits on a single semiconductor chip. It provided a considerable reduction in space as well as lower manufacturing costs. Significant development was reflected in larger memory capability for computers.

958. Integrated Circuits. The first read-only memory (ROM) for computers was brought out by Fairchild this year. These were 64 bit metal oxide semiconductor types.

LASERS

959. Lasers. North American Aviation Inc. developed a carbon dioxide laser having 4,000 watts output.

960. Lasers. A number of tests run by the United States
   and France using laser light from the ground, which
   was reflected from satellites, was claimed to have
   permitted the location of points on the earth to an
   accuracy within 10 cm.

PHYSICS

961. H.J. Prager, K.K.N. Chang and S. Weisbrod. A second mode of avalanche diode operation called the
   anomalous mode was discovered this year by Prager,
   Chang, and Weisbrod. In this mode, the operation
   was at a much lower frequency than the transit-time
   frequency. This mode became known as the
   TRAPATT mode, standing for Trapped Plasma Avalanche Triggered Transit.

962. Particle Accelerator. On October 9, the world's most
   powerful particle accelerator was put in operation at
   Serpukhov, USSR. An energy level of 76 billion
   electron volts shortly after start-up was claimed.

POWER

963. Nuclear Power. On June 14, the nuclear power station
   of Southern California Edison at San Onofre came into service. This power plant capacity was 430 MW,
   making it the largest commercial nuclear plant in
   the United States up to that time.

964. Krasnoyarsk Dam. The world's largest hydroelectric
   power plant was being constructed on the Yenisei
   River in Siberia. The plant, when put in operation,
   would provide over 6,000 megawatts--almost three
   times that of the Grand Coulee Dam.

965. High-Temperature Gas-Cooled Reactor. The first
   atomic power station using the high-temperature
   gas-cooled reactor (HTGR) was activated at the
   Peach Bottom Station in Pennsylvania. The station
   was rated at 40,000 kw and achieved a thermal efficiency of 37 percent. The HTGR system was developed by G.A. Technologies of San Diego, California.

RADAR

966. Underground Radar. Starting this year, the Electro-
Science Laboratory at the Ohio State University began research and development of underground radar
systems. The Terrascan, a battery-powered portable
unit, was developed for locating buried plastic and
metal pipes. Tests showed that plastic and metal
pipes could be located at depths of up to ten feet,
to within one foot and to within 15°. Penetration
was determined by the ground conditions.

RADIO

967. IC Radio. What is believed to have been the world's
first integrated-circuit radio was developed by the
Sony Corporation and put on the market in March.
It was known as the ICR 100 and was smaller than
a package of cigarettes.

RADIO ASTRONOMY

968. Robert Hjellming and C.M. Wade. The National Science Foundation's new 328-foot radio telescope at
the National Radio Astronomy Observatory in Green
Bank, West Virginia was put in operation this year.
Before the year was out, C.M. Wade and Robert
Hjellming had detected another star emitting electromagnetic energy. This was a nova, apparently in
the constellation Delphinus.

RECORDERS

969. Electronic Video Recorders. A new recorder was demonstrated by the CBS Laboratories known as the
EVR (Electronic Video Recorder). The device recorded picture and sound data on a tape that permitted playback into a standard home television set.

970. Video Recorders. The Ampex Corporation brought out
their videodisk recorder this year. The recorder
sold for $5,000.

SATELLITES

971. <u>ATS-III</u>. The third Applications Technology Satellite
   (ATS), built by Hughes Aircraft Company to provide technical information valuable for the future
   space effort, was launched on November 5 from
   Cape Kennedy, and went into synchronous orbit.
   This was the first satellite to send color photographs
   of the earth's full disc.

972. <u>Intelsat 2F-2</u>. A second attempt to put an <u>Intelsat</u> in
   a stationary orbit over the Pacific was made by the
   Communications Satellite Corporation. <u>Intelsat 2F-2</u>
   was launched January 11 from Cape Kennedy by a
   <u>Thor-Delta</u> rocket. It was put in position over the
   International Date Line to provide communication to
   Hawaii and Asia. Commercial service was started on
   January 27.

973. <u>Intelsat 2F-3</u>. Launched from Cape Kennedy on March
   23, <u>Intelsat 2F-3</u> was put in stationary orbit to sup-
   plement the Pacific area coverage. It was positioned
   over longitude 5° west. Service was started in April
   with 240 two-way voice channels.

974. <u>Vela Project</u>. Two more <u>Vela</u> satellites were put in orbit with a 70,000-mile altitude, approximately circular
   on opposite sides of the earth. These were to continue to check on any illegal nuclear tests.

TELEPHONE

975. <u>Telephone Ceremony</u>. On May 11, a ceremony marking
   the installation of the 100-millionth telephone in the
   United States was led by President Johnson. By
   this date over 90 percent of the telephones were
   equipped for direct dialing. Calls could be placed
   to 196 countries and territories. Direct dialing by
   the operators could be made to 23 countries.

976. <u>Submarine Cable</u>. Bell Telephone placed in service a
   coaxial cable system between Miami, Florida and
   Washington, DC carrying 32,400 conversations over
   nine cable pairs with two cables for spares. These

cables used the new L4 system which was introduced this year. The capacity was increased to 3,600 channels with repeaters placed at two-mile intervals.

TELEVISION

977. <u>International TV</u>. Television became international with the first global hookup in history arranged by the National Broadcasting Company. A two-hour international television program was aired on June 25. Nineteen countries on five continents participated with live programs. By means of four satellites, the program was viewed in 39 countries.

978. <u>European TV</u>. The TV stations in Europe began color TV transmissions this year.

979. <u>US TV</u>. In the United States, full-time color casting was used on ABC, NBC, and prime-time shows.

TRAFFIC CONTROL

980. <u>Traffic Control</u>. On-line computer ramp control of traffic was started in the Detroit area with manually controlled systems being discontinued.

<u>1968</u>

BROADCASTING

981. <u>AM Licensing Hold</u>. Applications for AM broadcasting stations (licenses) were accepted by the FCC up to July. At that time a halt was called to permit a study to be made on how to provide more noninterfering channels. At this time there were over 4,200 stations on the air, broadcasting on the 107 available channels.

COMMUNICATION

982. <u>Dataphone 50</u>. An updated version of Dataphone known as Dataphone 50 was introduced by the Bell Systems. This was a high-speed system of data transmission for computers and facsimile.

983. <u>Carterfone Decision</u>. The Carterfone was a device made to interconnect radio communication equipment to the telephone system by a base station. The device was manufactured by the Carter Electronics Corporation in Dallas, Texas. The Bell Company would disconnect such attachments or terminate service at that point. In June, the FCC ruled that the Carterfone could be connected to the Bell lines, and that customer-owned equipment could be attached.

COMPONENTS

984. <u>Light-Emitting Diodes</u>. The first commercially available light-emitting diodes (LEDs) appeared in 1968, an outgrowth of laser development. Several colors were available. Hewlett-Packard put out the first commercially available LED matrix, a 28-diode unit.

985. <u>Burndy Corporation</u>. The Burndy Corporation of Norwalk, Connecticut developed a connector using a high-pressure spring behind a pointed pin to force its way through any tarnish or corrosion on contact. A gas-tight cover provided a reliable and permanent connection.

COMPUTERS

986. <u>Computer Control</u>. Electronic Computers were used on the <u>Gemini 11</u> and <u>12</u> spacecraft to automatically perform the precise functions to return the space vehicles safely to earth.

INDUSTRY

987. <u>Edson D. deCastro, Henry Burkhardt III, and Roland</u>

G. Sogge. In April, the Data General Corporation was organized by deCastro, Burkhardt, and Sogge. The organization soon became well-known in the computer industry. Before the year was out, it had produced the NOVA, a computer selling for about $8,000.

INSTRUMENTATION

988. Logimetrics. Logimetrics brought out the 900 series signal generators. The units contained a built-in frequency counter which made exact frequency settings possible.

INTEGRATED CIRCUITS

989. ICs. By the end of the year, integrated circuits were taking over more and more of the functions of transistors and vacuum tubes and in about three years had replaced all but the kinescope in TV sets.

990. Integrated Circuits. RCA introduced the complementary metal oxide semiconductor (C-MOS) logic series, which soon made them a strong competitor of Texas Instruments and Motorola.

LASERS

991. Galium Arsenide Lasers. The most efficient solid-state lasers built to date were developed by RCA from crystals of gallium arsenide.

992. Laser. A Q-switched laser, having 10^9 w output, was developed at Compagnie Genérale d'Electricité of Paris.

MANUFACTURING

993. Computer-Aided Design. Computer-aided design and manufacturing began to be developed about this time. This technique permitted circuits or mechanical-

design performance to be checked, altered, and re-
checked strictly by computer operation. This process could be continued until the design was perfected and a tape made for automatic machinery control.

MEDICINE

994. <u>Artificial Limb</u>. An artificial arm that enabled the
above-the-arm amputee to mentally control the limb
was demonstrated at the Massachusetts General Hos-
pital. The limb sensed the electrical impulses generated by the mind when arm action was desired.
These impulses controlled electronic circuits which
drove the arm and hands.

PHYSICS

995. <u>J.L. Ferguson, G.H. Heilmeier</u>. The use of liquid
displays and their implementation is considered to
date from about this time. The work of Heilmeier's
group was described in the July issue of the <u>Pro-
ceedings of the IEEE</u>. Ferguson was awarded Patent
#3,410,999 (November 12) for a display system util-
izing liquid crystal material.

996. <u>Dr. Luis Walter Alvarez</u>. Dr. Luis Alvarez, known as
the Father of Ground Control Approach Radar, while
at the Radiation Laboratory of MIT, was awarded
the Nobel Prize in physics for the development.

POWER

997. <u>NERVA I Project</u>. A test reactor developed for the
Nuclear Engine for Rocket Vehicle Application Project (NERVA) developed 4,200 MW for 12 minutes at
the Rocket Development Station at Jackass Flats,
Nevada. This project was a joint AEC and NASA
effort.

RADIO

998. <u>Statistics</u>. This year, radio made the biggest annual
sales gain since the start of television broadcasts.
Income for radio exceeded $1 billion for the first
time in its history. Radio advertising had increased
a greater percentage than either newspapers or
television.

999. <u>Moss and Okonsky</u>. There were now more than 2,000
FM stations. Senator Frank Moss and Representa-
tive Alvin O'Konsky reintroduced in Congress the
All-Channel Radio Bill which required that all new
radios be capable of receiving either AM or FM signals.

RADIO ASTRONOMY

1000. <u>Antony Hewish</u>. British scientists at the Mullard
Radio Astronomy Observatory of Cambridge discovered what appeared to be radio signals generated
by an intelligent source. These were short pulses
of energy repeating at 1.337-second intervals,
originating in a region with no conspicuous objects
visible to astronomers. The team was headed by
Antony Hewish. Later three more signals were discovered. The term "pulsar" had been applied to
these pulsating radio sources. No pulsar has been
definitely identified with the location of a visible
star or galaxy. No satisfactory explanation was
known at this time.

RECEIVERS

1001. <u>A.L. Lovell</u>. Development of the scanning receiver
was attributed to A.L. Lovell of Regency about
this time. Scanning permitted the receiver to automatically switch in sequence through its channel
range, stopping when a received signal was detected. The first receiver was called the Bearcat.

TELEPHONE

1002. **Type SF Cable**. The type-SF ocean cable was installed across the Atlantic to link the United States
and Europe. The cable provided 845 two-way simultaneous voice channels. It was the first cable
using transistorized amplifiers. The repeaters
were spaced at ten-nautical-mile intervals.

TELEVISION

1003. **TK44A Camera**. RCA introduced the TK44A TV camera, capable of picking up acceptable color pictures
with only 15-foot candles of light.

1004. **X-ray Radiation**. A bill was passed by the House of
Representatives authorizing the Department of
Health, Education and Welfare to enforce the radiation standards after it was shown that many television sets, particularly color sets, exceeded the
safety standards.

1005. **Liquid Crystal Television Display**. The possibility of
using liquid crystal display for television had been
under study at the David Sarnoff Research Center
in Princeton, New Jersey. The results up to this
time indicated the feasibility of reflective television
display. Resolution of 150 to 175 lines had been
obtained.

1006. **High-Definition Television**. About this time NHK
Broadcasting Corp. in Japan started development
of a high-definition television system. Picture
resolution eventually exceeded 1,000 lines per
frame.

1007. **Hand-Held Camera**. The smallest TV camera ever
built was developed by RCA for the Apollo space
program. This was designed for use on the Apollo
7 flight. Small enough to be held in the hand, it
sent back the first live pictures from a US spacecraft.

1008. **High-Power TV**. The world's most powerful television station went on the air in Philadelphia this year. The output power was rated 110 kw. With its high-gain antenna, it had an effective radiated power of 4.3 MW.

1009. Canadian Television. Feeling the need for more television channels, the Canadian government ordered that all imported and Canadian manufactured television sets must be capable of receiving both UHF and VHF channels.

1010. 3-D Television. The possibility of three-dimensional television pictures, using lasers and holographic techniques, was demonstrated by several organizations this year.

1969

BROADCASTING

1011. Broadcast Statistics. By the end of 1969 there were nearly 4,300 AM stations and over 2,400 FM commercial stations in operation. The number of AM stations had reached the saturation point. The FCC was forced to prohibit the issuance of AM station construction permits except under special and unusual circumstances.

COMMUNICATION

1012. FCC Ruling. The FCC ordered AT&T to allow the interconnection of equipment owned by the customer to the public telephone lines. This action was the result of the Carterfone decision.

1013. Microwave Communication, Inc. Microwave Communication, Inc. was awarded a license by the FCC to set up a microwave system between Chicago and St. Louis in competition with AT&T.

COMPONENTS

1014. <u>Kim and Armstrong</u>. The IMPATT diode was improved again this year by C. Kim and L. Armstrong by a heat sink so as to lower the thermal resistance of the device, permitting 500 mw output at over 11 percent efficiency.

1015. <u>Transformers</u>. Ferrite cores for transformers and choke coils came into use about this time. The use of ferrites permitted inductors, in some cases, to be reduced to 1/100 to 1/1000 the size required when using the conventional cores.

1016. <u>Tubes</u>. Eimac brought out two new water-cooled tetrodes--the 50-kw 4CW50,000E and the 4CW100,000E.

1017. <u>Relays</u>. Miniature relays, no larger than the TO-5 cased transistors, were put on the market by C.F. Clair Co. This was a case approximately .370" (diameter) x .250" (height).

1018. <u>Color Tube for TV</u>. A new color tube had been developed by RCA which provided about twice the brightness in the TV picture as the conventional color tube. The tube provided better color when viewed under normal light conditions.

COMPUTERS

1019. <u>IBM 195 Computer</u>. IBM introduced a computer known as the Model 195 which was said to have been able to process an instruction in 54-billionths of a second.

FABRICATION TECHNIQUES

1020. <u>Ion Implantation</u>. It was about this time that the technique of ion implantation in silicon surfaces came into use in the production of integrated circuits.

INDUSTRY

1021. **RCA Corporation**. In a program stated to modernize
the company's identity, the corporate name was
changed from Radio Corporation of America to RCA
Corporation.

INSTRUMENTATION

1022. **High-Speed Oscilloscope**. A new oscilloscope was announced by Hewlett-Packard Co. near the end of
the year, having a response from DC to 250 MHz
yet useable at least to 500 MHz. The writing speed
was 4 ns/cm. Known as the model 183, it was recognized as the fastest direct-writing oscilloscope
up to this time. The instrument was all solid state
except the cathode-ray tube.

LASERS

1023. **Laser Weapons**. The development of high-power lasers
led to experiments for their use as military weapons.
Tests at Kirtland Air Force Base, New Mexico resulted in a drone aircraft being shot down by a
carbon-dioxide laser.

1024. **Magnetically Tuned Laser**. A magnetically tuned spin-
flip Raman infrared laser was developed at Bell. It
had valuable applications for high-resolution spec-
troscopy.

1025. **Neil Armstrong, Michael Collins, and Edwin E. Aldrin**.
An experiment to determine the exact distance from
a point on the earth to a point on the moon was desired as a means of determining the answers to
some of the questions about the earth such as the
continental drift. For this purpose, a reflector
was set up near the lunar landing site of the Columbia spacecraft on the moon. The reflect was
set up by Armstrong and Aldrin while Collins acted
as pilot of the Apollo 11 craft orbiting the moon.
Laser returns were received by both the McDonald
and Lick Observatories in Texas. By using laser

pulses of only a few billionths of a second, long-distance measurements could be made to within a few inches.

MEDICINE

1026. Nuclear-Powered Pacemaker. In May, a pacemaker powered by a radioactive isotope was implanted in a dog to regulate the heartbeat. It was said to emit no more radiation than a radium dial wrist-watch. If the device proved satisfactory, it would have a life of about ten years rather than the two-year life expectancy of battery-powered pacemakers.

POWER

1027. Statistics. By the end of the decade, the construc-tion of nuclear power stations was declining as the public objected to their construction near populated areas. Increasing construction costs also forced the termination of some projects. The total power produced by nuclear stations this year was about 1.443-trillion kwhr.

1028. Nuclear Power. The Oyster Creek (New Jersey) plant of the Jersey Central Power and Light Com-pany went critical on May 3. When brought up to full power, it was reported to have been the first nuclear plant to compete with fossil fuel in produc-tion costs. Its rated capacity was 640 million kw.

1029. Uranium U-233 Reactor. The first atomic reactor fueled by isotope U-233 was put in operation at Oak Ridge National Laboratory, Oak Ridge, Ten-nessee. The reaction of U-233 would breed addi-tional fuel by the capture of neutrons in thorium-232, thus reducing the cost of operation.

RADAR

1030. Ground-Controlled Approach Radar. Starting in Janu-ary, the Raytheon Company was given a contract

for the development of the AN/TPN-19 GCA Radar.
It was to be all solid state except for the RF power
amplifiers modulators, and cathode-ray tubes.

RADIO ASTRONOMY

1031. Pulsars. By the end of 1969, at least 40 pulsars had
been discovered. In January of 1961, the first
visible pulsar was detected in the Crab Nebula and
had the shortest pulsating period of any known
pulsar. It flashed at a rate of 30 times per second.
The star was discovered at the Steward Observatory
in Arizona. The discovery was confirmed by the
Lick Observatory in Texas.

RECORDING

1032. Quadraphonic Tapes. The first four-channel prerecorded tapes appeared on the market along with
playback equipment. Four-channel stereo programs
had been put on the air in Boston by stations
WGBH with WCRB cooperating. Later, WNYC and
WKCR in New York City also tried the four-channel
experiment. The idea was to give the impression
of a large auditorium with two speakers for the
conventional playback, and two speakers behind
the listeners, representing the reflections from the
concert hall. The first experiments were not as
satisfactory as were desired.

SUPERCONDUCTIVITY

1033. Superconductive Motor. The first large motor employ-
ing a superconducting winding was run in October
1969. By March 3, 1970, the motor was run at full
load--3250 HP at 200 RPM. The machine was designed and developed by the International Research
and Development Co., Ltd., of Newcastle, England.
It was named the Fawley motor after the site where
it was to be used.

TELEPHONE

1034. Boston-Miami Cable. The final link of the Boston-to-
       Miami telephone cable was completed this year. The
       new 20-tube cable systems handled over 32,000 si-
       multaneous conversations.

1035. President Nixon, Edwin Aldrin and Neil Armstrong.
       The longest long-distance telephone call was made
       on July 21. President Nixon in the White House
       talked to Neil Armstrong and Edwin Aldrin on the
       surface of the moon. Total distance covered on
       land lines and radio for the call was about 290,000
       miles one way.

1036. Automatic Telephone Exchanges. By this date, most
       telephone systems in the US had been equipped
       with automatic exchanges and conversions were be-
       ing made to automatic exchanges in many other na-
       tions.

1037. Telephone Statistics. In 1969 about 115 billion calls
       were handled by the Bell System. This was 10
       billion more than in 1968. By this time, about
       115.3 million telephones were in service in the US.

TELEVISION

1038. TV Statistics. By the end of 1969, an estimated 862
       television stations were in operation with over 88
       million TV receivers in the United States and its
       possessions.

1039. Apollo 10. In May, Apollo 10 sent color pictures of
       the moon back to earth. On June 21, pictures of
       the Apollo 11 crew walking on the moon were sent
       to earth.

1040. Remote-Controlled TV. RCA introduced a remotely
       controlled TV color set, tuned by small ultrasonic
       transmitters.

TIME

1041. <u>Japanese Wristwatch</u>. Japan brought out its first
quartz watch in late 1969. This watch used a 14-
stage frequency divider and a hybrid-type circuit
assembly.

TRANSPORTATION

1042. <u>Magnetic Levitation</u>. The experimental use of magnetic levitation for public transportation was started
in Germany when a study of a high-speed train system was started under the direction of the Bonn
government. An experimental system (unmanned)
travelled at over 240 mph.

Chapter 4

THE EIGHTH DECADE: 1970-1979

By 1970, the space age had matured and was in full swing. Other nations besides the United States and the USSR had satellites in space. By the end of the decade there were hundreds of man-made objects orbiting the earth and providing various services to nearly all the major nations of the earth. At least six nations had satellites in orbit for various purposes and studies. There were satellites for communications, weather reporting, meteorite hazard, air drag, gravity, radiation, cosmic ray, aurora, and many other scientific studies. In the 1970's many hundreds more joined those already in orbit. A good many of these were communications satellites which were providing communication channels to at least 80 countries by 1971. Some of these provided television and various forms of data transmission. Coast-to-coast telephone service via satellite was producing good service with considerable savings to the public. By the middle of the decade continuous cloud-cover pictures of the United States were being returned to the surface to assist weather forecasters. Other satellites provided improved communication to ships at sea.

Besides the many satellites providing various services for many countries, the US and the USSR launched a number of space probes, which returned thousands of pictures to Earth from various areas of the solar system. Close-up pictures of the planets and their moons from Mercury to Jupiter and beyond were made with soft landings on Venus and Mars giving surface pictures of amazing detail.

Due to the satellites and extended cable systems around the world, by 1978 at least 124 countries had direct telephone service to the US. Still other improvements to the telephone system were being made. Picturephone systems were being

installed in some of the larger cities. When the cost of the
system was reduced, the interest in picturephones expanded.
International direct-dialing was coming into being with the
first extension of the service to Britain. Automatic-intercept
electronic switching and improved instruments further improved telephone service during this period. Communication
secrecy and tamper-proof telephone service were also considerably advanced by improvements in light-guide fibers and
connectors for them. Experimental light-guide service was
tried in some areas.

The expansion of television service followed that of the
telephone. By the opening of the decade, NBC programs
were being seen in 114 countries. The first China-US telecast was made, showing President Nixon at the Great Wall.

The development of smaller components along with great
progress in integrated circuits permitted small portable television receivers to be produced. The first had a 4 1/2-inch
diagonal screen, still later a smaller set with a 2-inch screen
was produced. Research continued on still smaller sets.
The small size of the television screens had been made possible by the development of liquid crystal screens. These
did not require the depth or voltage of the conventional
television picture tubes.

The same technique used to produce the small television
screens was also applied to the electronic watches during
this period. The liquid crystals made ideal watch readout
plates. With the development of large-scale integration chips,
the circuitry was so reduced in size that multifunction wrist
watches came on the market. Within a few years the prices
of good reliable watches were reduced to the order of $15.
Before the end of the decade, a solar-powered watch was on
the market.

During this period the nations of the world adopted a
new time system known as Universal Coordinated Time (UTC).
Until January 1972, several time systems were in use in various countries. Initially (before 1955), time was determined
by astronomical observation. By international agreement the
meridian of Greenwich, England was taken as the standard
meridian for both longitude and time. It had been recognized for some years that time derived from earth's rotation
would vary slightly because of unpredictable and long-term

variations in the rotation of the earth. A more accurate time scale was required for some scientific measurements.

Ephemeris time, that is, time based on the revolution of the earth around the sun, is free from earth-rotational error and was introduced in 1950. Because of the long time required for checks, it was not suitable or convenient for some scientific applications. With the development of the atomic time standard a constant-length second was now possible. By international agreement in 1971, the basic time was to become based on an atomic standard effective January 1, 1972. This was the official start of Universal Coordinate Time. Occasionally "leap-second" corrections are made to prevent a wide drift in time relative to ephemeris time.

New developments in integrated circuits and fabrication techniques had resulted in great reductions in cost as well as in the size of chips. Early in the decade, pocket-sized calculators came on the market at a price affordable to nearly anyone needing them. These calculators provided capabilities exceeding the requirements of most users. As a result, simpler calculators which were primarily suitable for arithmetic computing came on the market. As the demand increased, the cost decreased. It was not long before small calculators became competitive in price with the slide rule. The slide rule was almost obsolete for arithmetic computing.

Before the end of the decade, calculators operating on solar or available ambient light were brought out. The most obvious example of the advantage of integrated circuits was in the computer area. The development of the very small computers, condensing a room full of equipment to a volume hardly exceeding that of a suitcase and reducing the power of the early computers from kilowatts to watts, was quite amazing. Further component and circuit-size reduction brought the device into other fields such as toys and automobiles.

Another device permitting tremendously increased computer memory with minimum size was the holographic disc memory. It was scratch-resistant and permitted an extremely high density of memory on one disc, yet kept the short access time to any given address.

Other applications of the laser were being investigated. There was a reduction in size, resulting in the development of pocket-sized, battery-powered lasers. Laser power was also used in military applications. Several reports of drone aircraft being shot down by laser power were publicized.

Laser applications for welding came into use in some automobile factories. A new welding technique was announced --low-temperature welding. An electrical pulse of very high current for a very short period of time was applied to provide a reliable weld but with little excess energy to provide area heating.

Research on magnetic levitation was carried on, particularly in Japan and Germany, with the aim of developing high-speed transportation, rivaling that of airplanes.

The largest superconducting motor yet developed was implemented in England. The motor developed 3250 HP at 200 rpm.

The requirements for more efficient batteries, undoubtedly brought on by the requirements of spacecraft, satellites, watches and other battery-powered devices, resulted in a number of new types during this period. A number of cells with higher energy output than the normal cells were introduced. Other anticipated uses for batteries, such as those required for electric cars, were guiding some research. An interest in electric cars for short-haul use was being considered, primarily because of the increasing price and scarcity of fossil fuels.

1970

AMATEUR

1043. Oscar 5. The fifth Orbiting Satellite Carrying Amateur Radio (Oscar 5) was launched into a 115-mile polar orbit on January 23. It was the first Oscar under active ground control. Two-meter and ten-meter beacons were carried. The two-meter beacon operated 23 days and the ten-meter beacon operated 40 days. Power was provided by internal batteries.

1044. William Byrd and Paul Wilson. A world's record of
      249 miles for communication on 2300 MHz was estab-
      lished July 11 by William Byrd, WA4HGN of Muscle
      Shoals, Alabama and Paul Wilson W4HHK of Collier-
      ville, Tennessee. The power used was under one
      kilowatt.

1045. Paul Wilson and Bill Smith. After about three and a
      half years of effort, Paul Wilson, W4HHK, and Bill
      Smith, W3GKP, succeeded in communicating by moon
      bounce on 2300 MHz. On October 19 in a two-way
      exchange, a new record was made with the highest
      frequency ever used for earth-moon-earth communi-
      cation.

      COMPONENTS

1046. Photovoltaic Cells. The top efficiency of photovoltaic
      cells at this time was about 14 percent.

1047. Picture Tube. By increasing the deflection angle
      from 90° to 110°, Sony was able to reduce the
      length of their 18-inch picture tube by four inches,
      providing a considerable reduction in volume of
      their television sets.

1048. IMPATT Diode. The IMPATT diode, discovered at
      Bell Laboratories in the mid-1960's as being a mi-
      crowave generator, was so improved by this time
      that over four watts of power could be produced
      at five gigahertz with an efficiency of about 13
      percent.

1049. Optical Isolators. A line of optical isolators, or coup-
      lers as they are sometimes called, was put out in
      Dual In-line Packages by Monsanto. These devices
      provided extremely high isolation with low noise.

1050. Tubes. The development of new tubes had been dis-
      continued by most companies. The development of
      transmitting tubes had been continuing with two
      new lines out this year by Eimac. These were a
      new line of tetrodes which included the 4CX600B,
      4CX600F, 4CW800B, 4CW800F and the 4CX600J/8809.

A second group of planar triodes included Y503, Y518, 7815AL, and 8847.

1051. Integrated Circuits. A major breakthrough for computers was made by Intel with the introduction of a 1024-bit MOS Random Access Memory at a reasonable price. By this time, integrated circuits were being perfected and production quantities were giving high yields; as a result, with decreasing prices, computer use was spreading to other than the electronic field. Within a few years, integrated circuits were being used in toys, air conditioners, furnace controls, automobiles, telephones and many other devices.

1052. Fiber Optics Attenuation. Starting in the 1960's with fiber optic losses on the order of thousands of decibels per kilometer, losses had continually come down to where, by 1970, attenuations on the order of 20 db/km were about average. Development on fiber optics was continuing and by 1973 losses of 2 db/km were obtained at 1 μm wavelength.

1053. IBM Mini-Computers. IBM entered the mini-computer field with the introduction of the model 145 of its system 370 line. This was the world's first large, general-purpose computer to use monolithic semiconductor circuits for the entire main memory.

1054. Data General Corp. Data General Corporation became one of the first manufacturers to offer a line of small computers using metal oxide semiconductor large-scale integration.

INDUSTRY

1055. General Electric Co. General Electric, disillusioned with the time-sharing technique it had developed for computers, sold its computer business and assets to Honeywell Information Systems.

1056. General Radio Company. In March, the General Radio Co. purchased controlling interest in Time/Data Corporation of Palo Alto, California. In four years,

Time/Data had become a recognized leader in the field of digital signal analysis. A second expansion was made about a month later when an agreement was made by Grason-Stadler and the General Radio Co. Grason-Stadler, a recognized leader in the manufacture of instruments for the life sciences, became a wholly owned subsidiary of General Radio.

INSTRUMENTATION

1057. **Dr. Albert Crewe.** Dr. Albert Crewe of the University of Chicago received the industrial research award as "Man of the Year" for his technique of using holography to visualize single atoms with the electron microscope.

LASERS

1058. **C.K.N. Patel.** C.K.N. Patel of the Bell Telephone Laboratories produced the high-power carbon-dioxide C.W. laser.

1059. **Pocket-Size Laser.** A battery operated laser, small enough to fit in a coat pocket, was developed.

MEDICAL

1060. **Nuclear-Powered Pacemaker.** The first known implantation of a nuclear-powered pacemaker was done in France this year.

1061. **Particle Accelerators.** The world's most powerful Van de Graaff nuclear-particle accelerators were put in operation at the Brookhaven National Laboratory on Long Island. Proton acceleration up to an energy of 30 MeV had been achieved.

POWER

1062. **Reactor Statistics.** In 1970 about 476 land-based nuclear reactors were in operation around the world,

109 of which were power reactors; 367 were research reactors.

RECORDING

1063. Four-Channel Tapes. RCA began production of four-
channel tape recordings on Quad-8 cartridges.
Wollensak and Sony brought out open-reel tape
decks.

TELEPHONE

1064. Picturephone. Picturephone service was first introduced to commercial service in Pittsburgh, Pennsylvania in July by the Bell Telephone Company.
It was not widely accepted at that time, partly because of the cost of $165 per month. By 1971 the
cost had been reduced to $25 per month. At least
100,000 of these phones were expected to be in
use within five years.

1065. Automatic Intercept System. An automatic system to
let customers know how to complete calls that reach
nonworking numbers was announced by the Bell
System. Installation was started in selected areas
this year.

1066. International Telephone Dialing. The international direct distance dialing (IDDD) was started in March
with direct dialing between New York City and
Great Britain. Initially 24,000 customers in New
York could use the system.

1067. Telephone Statistics. As of 1970 the United States
had 115.2 million telephones in service. This was
about 45 percent of the world's total.

1068. Telephone Cable. On March 22, the highest capacity
communication cable ever laid across the Atlantic,
called TAT-5, was put in operation linking Green
Hill, Rhode Island and San Fernando, Spain. The
additional cable led to international rate reductions
for transatlantic telephone calls, some by as much

as 25 percent. The cable was jointly owned by AT&T and ITT World Communications, Western Union International, RCA Global Communications and some European Organizations.

1069. Telephone. The French electronic switching for telephone communication, known as E10, was introduced this year. This system could handle 90,000 calls per hour. The system utilized digitally encoded signals and time-division multiplexing.

TELEVISION

1070. National Broadcasting Co. By the end of the year, NBC was providing television programs to 114 countries.

1071. Television. The first truly portable color television was announced in 1970. It was manufactured under the name of Panasonic. The color tube was a 4-1/2-inch diagonal. The set required 12 volts from a 2.6-pound battery, self-contained. The set also had an AM-FM radio.

TIME PIECES

1072. Time Pieces. An alarm was applied to the electronic watch. The idea was based on electromagnetic glass excitation.

1971

COMPONENTS

1073. Batteries. Sealed lead-acid batteries were brought out by the Gates Rubber Company of Denver, Colorado. These batteries provided the highest watts-hours per dollar available. The cells were sealed in a leakproof steel case needing no additional water. The cell had very low internal resistance

and discharge rates as high as 100 amps could be
obtained from a D cell.

1074. ZIF Connectors. With the development of large-scale
integration, more and more terminals were being
required for external leads. This became an increasing problem in inserting or removing integrated
circuit boards because of bent or broken terminals.
About the middle of the year, Textool Products of
Irving, Texas brought out their ZIF connector.
This zero insertion force (ZIF) socket locked the
pins after insertion, forming a good, noise-free
connection.

1075. Magnetoresistor. The Siemens Corporation of Iselin,
New Jersey announced a new device known as the
"Magnetoresistor." This was a resistor element
which changed its resistance with the magnetic field
around it. Variation in resistance from 12 to 18
times could be obtained by magnetic field variation.

1076. Gas Sensing Semiconductor. Figaro Engineering Inc.
of Kobe, Japan developed a semiconducting material
composed of metals such as oxides of tin and zinc
and forms of iron oxide that decreased its resistance
in such gases as hydrogen, carbon monoxide, pro-
pane and alcohol. Conductivity returned to normal
when the gas was removed.

1077. Silicon-Controlled Rectifiers. General Electric introduced the first silicon-controlled rectifier in 1957.
By 1971 it had manufactured about 50 million SCRs
for use in appliances, business machines, automobiles, industrial process control and other uses.
They are now used in circuits up to 2,600 volts
and for currents as high as 1,300 amps.

1078. High-Frequency Transistor. The IBM Laboratories
in Zurich, Switzerland developed a transistor with
a cut-off frequency of 30 GHz. It provided a 3-db
gain at 17 GHz--more than twice that of any known
transistor up to this time.

1079. IMPATT Diodes. By this year, the IMPATT diode,
originally developed at the Bell Telephone Labora-

tories in the mid-1960's, had become the prime solid-
state device for generating microwave power up to
300 GHz.

1080. Tubes. Eimac brought out a new group of transmitting tubes: 8873, 8874, 8875, 8877, 8121W, 8972W,
8828W, 8122W, and X2159.

COMPUTERS

1081. Microcomputer. Early in the year, the Mostek Corporation of Dallas, Texas announced it had developed a large-scale integration (LSI) circuit containing complete logic for an electronic calculator.
The single chip replaced 22 medium-scale integration
circuits. The chip made possible a computer about
the size of a cigarette package.
Almost immediately, Texas Instruments disclosed
that it, too, was developing a one-chip calculator
that would be out by June. The one-chip computers became known as microcomputers.

1082. Niklaus Wirth. The computer language PASCAL was
invented in Zurich, Switzerland by Niklaus Wirth.
This computer language was named after Blaise
Pascal, the French mathematician.

INDUSTRY

1083. Potter-Englewood. The firm of Potter-Englewood acquired the assets of the Muter Company and the
new company incorporated under the name of Pem-
cor, the acronym for Potter-Englewood-Muter Corporation.

1084. Radio Corporation of America. The Radio Corporation
of America went out of the general-purpose computer business and sold its computer assets to the
Univac Division of Sperry-Rand Corporation. RCA
continued to develop data communication systems.

INSTRUMENTATION

1085. Speech Speed-Up. A solid-state device, approximately one cubic inch in volume which permitted recorder manufacturers to build in variable-speed speech control, was developed. The module permitted playback of normal speech at either a lower or higher speed than the original without altering the pitch. The speech was said to have been clearly understood at speeds up to at least four times normal. The device was patented by the Cambridge Research and Development Group of Westport, Connecticut.

1086. Night-Viewer. GTE Sylvania announced their new night-viewer, developed primarily for police and security officers. The device is reported to detect man-sized objects to about one third of a mile in moonlight. It contained a light amplifier with a light intensification of up to 45,000 times.

INTEGRATED CIRCUITS

1087. Isoplaner Process. The Isoplaner process of fabricating integrated circuits was developed at Fairchild. The new process permitted a further reduction in the size of the circuit chips.

1088. Solid-State TV Circuits. The first integrated circuits and solid-state techniques were being applied to new television sets brought out this year. The advantage claimed was higher reliability for the set.

LASERS

1089. Color Laser. The RCA Electronics Components Division demonstrated a helium-selenium laser which produced 24 separate colors. The disclosure was first made in February at the American Physical Society Conference in New York City.

MEDICAL

1090. Nuclear-Powered Pacemaker. The first request in the US for the implantation of nuclear-powered pacemakers was filed this year with the Atomic Energy Commission on behalf of ten patients. The devices were powered by plutonium-238. The pacemaker was designed by a French company, and the first implantation in a US citizen was made in 1972 in Buffalo, New York.

PATENTS

1091. Michael Cochran and Gary Boone. The basic patent on the single-chip microcomputer was filed by and later awarded to Michael Cochran and Gary Boone of Texas Instruments, Inc.

PHYSICS

1092. Dennis Gabor. Dr. Dennis Gabor was awarded the Nobel Prize in physics for his invention of holography. It was not until the laser was invented that a suitable source of coherent light was available to make holography practicable.

POWER

1093. Atomic Power Statistics. By the end of 1971, 22 nuclear power plants were in operation in the United States.

1094. Hydroelectric Power. This year, the world's largest hydroelectric power plant was put in operation in Krasnoyarsk, Russia. The plant had a reported output of 6,000 MW.

SATELLITES

1095. Intelsat 4F-3. Intelsat 4F-3 was launched from Cape Kennedy by an Atlas-Centaur rocket on December

19. This satellite contained a much sharper beam than previously, covering only a small portion of the earth. The resulting effective power increase permitted about 4,000 communication circuits and channel capacity was limited by bandwidth rather than power. At this time, 80 countries were members of INTELSAT.

1096. SATCOM. On January 8, RCA announced the satellite SATCOM. This was a communication satellite which provided coast-to-coast communication at an appreciable savings to the user over the existing system.

1097. Mariner 9. On May 30, the United States launched the satellite Mariner 9 towards Mars. The purpose of the probe was to map the surface of Mars, study its features and atmosphere and obtain photos of Phobos and Deimos, the two martian moons. On November 13, Mariner 9 orbited Mars within about 870 miles of the surface. A dust storm obscured the planet's surface at that time. Dust began to settle after about a week in orbit and the satellite began to take useful pictures. Summer temperature was about -80° F. Photographs of the moons showed that they were irregular in shape and scarred by many craters. Photographs and data were returned for about a year before it was shut off in October 1972.

SUPERCONDUCTIVITY

1098. Fawley Motor. In March, the superconducting motor installed at the Fawley site in England was brought up to full load. It produced 3250 HP at 200 RPM.

TELEVISION

1099. James H. Meacham. The smallest television camera ever built up to this time was disclosed at the National Aerospace Electronics Conference in Dayton, Ohio. The design was carried out under the direction of James H. Meacham of the Westinghouse De-

fense and Space Center, Baltimore, Maryland. The
camera used a 1/2-inch diameter Vidicon image
tube.

TIME

1100. Time. Integrated circuits and large-scale integration
programs had, by this year, made electronic
watches possible. By the end of the year, electronic watches were on the market selling from
around $275 to $2,000. In several years, watches
had been reduced in price about 50 percent or
more. The first electronic digital watch, brought
out in 1971 by Time Computer, was the Pulsar.
The time was shown by light-emitting diodes when
a button was pressed. The day and date readouts
were added in 1972.

WELDING

1101. Welding. Developed this year were high-speed low-
temperature techniques of welding that produced
strong welds so rapidly that nearby painted metal
was not destroyed. Known as "ultrapulse," it applied approximately ten times the normal welding
current through the metal but only for a fraction
of a second, not giving the metal time to heat up a
wide area. The system was developed jointly by
Quanta Welding Co. of Troy, Michigan and the
Linde Division of the Union Carbide Corp. in
Moorestown, New Jersey.

1972

AMATEUR

1102. Oscar 6. The Oscar 6 satellite was launched October
15 aboard the Itos-D rocket and put in a 900-mile,
115-minute circular orbit. A number of new records
were set by this satellite. Oscar 6 separated from

the launch vehicle over Eastern Africa. This was an active-repeater satellite which could be programmed on command from the ground. It could also store messages to be repeated later on command. Another first for the amateur was made when W9NTP and WA9UHV exchanged slow-scan television pictures via Oscar 6.

COMMUNICATIONS

1103. Data Communications. By this time AT&T was providing approximately 75 different types of data communications with about 20 speeds of operation. A direct-dialing network handled digital data at speeds of 4,800 bits per second with private-grade channels operating at speeds up to 10,800 bits per second.

COMPONENTS

1104. Batteries. The Bell Telephone Company put in use a new type lead-acid battery to provide standby power for central station operation. The new battery used pure lead grids. The life span of the new cells was at least twice that of the cells previously used. A life of over 30 years was claimed for the new cells.

1105. Picture Tubes. Picture-tube depth had again been reduced by Sony by increasing the beam deflection angle to 114°. Later in the year the angle was increased to 118° for the 16" tube.

1106. Light-Emitting Diodes. IBM Researchers produced light-emitting diodes (GaAs) that produced a 2-μm diameter beam. This was ideal for putting all the light into an optical fiber.

1108. Transistors. The Microwave Semiconductor Corp. announced the breakthrough in power transistors. The new five-watt cw 4-GHz transistor was announced in March. Power gain for the device was 4 db with an efficiency of 30 percent.

COMPUTERS

1109. Pocket Computer. The Hewlett-Packard HP-35, a
small, hand-held scientific calculator, was marketed
this year. The development of the computer was
started in 1970 by the Advanced Product Division
of Hewlett-Packard. The finished product weighed
nine ounces and could perform trigonometric, log-
arithmic, square root and other commonly used
computer functions. This was the first of many
competitive machines that appeared on the market.
By the end of the year about a half million of these
and similar devices made by a number of manufacturers had been sold.

GAMES

1110. Magnavox Consumer Electronics Company. An electronic game, designed to be played through a conventional television set by viewing the screen, was
brought out about the end of the year by Magnavox
Consumer Electronics Co. This was the start of a
new industry in electronics. Soon electronic games
were brought out by a number of companies. The
games eventually came out representing almost all
of the popular sports and games such as chess,
checkers and others.

INSTRUMENTATION

1111. Probeye. The Hughes Aircraft Company developed a
device called Probeye. This was a hand-held instrument used to permit night vision by infrared
rays.

INTEGRATED CIRCUITS

1112. Camera Control. The application of integrated circuits for camera control started as camera manufacturers applied automatic shutter control calculated
from the exposure meter indication.

1113. **EPROM.** The erasable programmable read-only memory (EPROM) integrated circuit was developed by Intel. The first unit was for 2,048 bits.

1114. **IC Applications.** This year the integrated circuits had become so cheap that toy companies and other manufacturers were bringing out a number of toys and electronic games using these electronic circuits.

MEASUREMENTS

1115. **Measurements.** The accuracy of the US legal volt at the National Bureau of Standards was now being maintained to a few parts in 10^8. This was achieved by a system known as the Josephson effect, which tied the standard to the invariant constants of nature.

PHYSICS

1116. **Proton Accelerator.** The proton accelerator of the National Accelerator Laboratory near Batavia, Illinois exceeded its design energy level of 200 billion electron volts, eventually reaching accelerations of up to 300 billion electron volts.

1117. **Mesons.** Los Alamos Accelerator at Clinton, P. Anderson Meson Physics Facility in Los Alamos, was tested at its design rating of 800 MeV. It is often called a meson factory because of the large number of mesons produced.

1118. **Bardeen, Cooper and Schrieffer.** John Bardeen, Leon N. Cooper and John R. Schrieffer, all US citizens, were awarded the Nobel Prize in physics for their theory of superconductivity. Bardeen was the first man to win a second Nobel Prize in a single field.

POWER

1119. **Nuclear Power Statistics.** As of June 30, 26 nuclear

power plants were in operation in the United States.
Fifty-one were under construction and 66 were on
order worldwide.

1120. Canada. The first full-scale commercial nuclear power
plant in Canada was dedicated on February 26.
The plant was located near Toronto.

SATELLITE

1121. Anik I. Canada's first domestic communication satellite was launched from Cape Kennedy, on November
9 and positioned 22,300 miles over the equator off
the Mexican coast. From this position it was able
to provide telephone, television and data service
to nearly all of the northern countries. The Canadian Broadcasting Corporation leased three channels
for television in the French language and others for
English-language television.

SPACE PROBE

1122. Pioneer 10. The Pioneer 10 spacecraft was launched
on a trajectory to Jupiter on March 2. On the trip
it returned information on micrometeoroid density,
solar wind, cosmic rays and magnetic fields in space.
It passed Jupiter in December 1973 at a distance of
about 80,000 miles. The craft was equipped with
four SNAP-19 (Space Nuclear Auxiliary Power) systems fueled by plutonium-238; each unit would produce 37 watts. The power units were expected to
last about six years. It was still returning information when it left the solar system over ten years
later.

1123. Venera 8. The USSR space probe Venera 8 was
launched on March 27. On July 22, the descent
package was released and landed on the sunny side
of Venus. Data from the surface were relayed back
for about 50 minutes before failing, apparently due
to the 878° surface temperature and the atmospheric
pressure (about 90 times that of sea level on earth).
The atmospheric analysis showed less than 0.1 percent oxygen, most being carbon dioxide.

TELEVISION

1124. <u>Charles H. Sewell-President Nixon</u>. The prediction
of Charles Sewell made in 1900 was realized in reverse order with President Nixon's trip to China.
In 1900, Sewell wrote:

> "...A child born today in New York City, when
> in middle age, shall visit China, may see repro-
> duced upon a screen, with all its movement and
> color, light and shade, a procession at that mo-
> ment passing along his own Broadway...."

TIME

1125. <u>Universal Coordinated Time</u>. The new UTC system
of time signals came into effect on January 1 in all
countries.

1126. <u>Motorola</u>. The first integrated circuit kits for watches
were sold to watch manufacturers for $15 by Moto-
rola.
Liquid crystal technology was being applied to
watches for readout display.

1127. <u>Leap Seconds</u>. In 1967, the official definition of the
second, based on solar time, was dropped, and a
new standard was adopted based on atomic time.
Solar time and atomic time did not track exactly,
due to earth wobble. On January 1, the use of
universal coordinated time and the "leap second"
was adopted. With this system, a "leap second"
would be added or subtracted as necessary when
the two times disagreed as much as one second.
The leap second was added or subtracted from the
last minute in June or the last minute of the year.

1973

AUTOMATION

1128. <u>Automation</u>. The world's first commercially available computer-controlled lathe was introduced in September by General Automation Inc. The lathe used was the American Tool Company's Hustler.

BROADCASTING

1129. Aerostat. Westinghouse was experimenting with television transmission from a tethered balloon over the Bahama Islands. Television programs from the United States were relayed by a balloon-borne transmitter.

1130. Broadcasting. After an eight-year freeze on new broadcasting station licensing, the FCC began again accepting and processing applications for new AM stations. By the end of 1973 there were about 4,500 AM radio stations in the US, 3,500 FM and 700 noncommercial educational stations.

1131. Citizens Band. This year citizens band operation became widely accepted by the public. Although CB operation had been legal since 1957, only about one million sets were in operation by 1973. In just two years, the number of users would increase another million sets. By the end of 1976, an estimated 11 million sets were sold.

COMPONENTS

1132. Batteries. The lithium battery came into production at the Mallory Battery Company. The new cell had a shelf life of up to 20 years. The operating range was up to 165° F and the cell provided an energy density of 200 watt-hours per pound, approximately twice that of previous cells.

1133. High-Power Transistor. A high-power transistor, having a peak collector-emitter voltage rating of 2,200 volts with a 2-ampere current rating, was developed by Texas Instruments, Inc. The transistor was designed primarily for the deflection circuits of television sets and had a switching time of 0.7 µs at 1.5 amperes.

1134. Oscar Heil. A new type of speaker was marketed this year. Invented by Oscar Heil, the speaker employed a polyethylene diaphragm folded in pleats between two permanent magnets. Conductive strips on each side of the folds provided the drive. The claimed advantages for the speaker were a low mass motion for a given output and high acoustical efficiency with nondirectional radiation. Initial promotion was for a mid-range tweeter.

1135. Lincoln Walsh. Another form of speaker was proposed by Lincoln Walsh. It appeared to be a conventional speaker, but the cone did not operate in the same way. The audio frequencies passed from the voice coil to the outer rim in concentric waveforms, transferring energy to the air. Two models were brought by the OHM Corporation this year.

1136. Shibaura Electric Co. The Shibaura Electric Co. in Tokyo, Japan came out with a short 16" TV tube. The deflection angle had been increased to 118°.

1137. IMPATT Diodes. Continued research on the IMPATT diode had improved efficiencies up to about 24 percent. Power up to about three watts continuous wave was obtained on the X and KU bands (5200-10,900 MHz and 15,350-17,250 MHz).

1138. Solid-State Tube Replacements. Electronic Devices, Inc. of Yonkers, New York brought out a line of five solid-state "tubes." The devices had essentially the same characteristics as the vacuum tubes they replaced. The five units could replace 25 tubes in use in radio sets, except no filament power was required. The life of these replacements was claimed to be about ten times the life of the tube replaced.

1139. Tubes. New tubes brought out this year by Eimac included the X-2170 and the X-2176.

COMPUTERS

1140. Speech-Synthesis. Bell Laboratories developed a com-

puter program to create artificial speech produced
from the printed word. The computer looked up
the word in memory and assigned duration pitch
and intensity to the word elements. The process
was demonstrated this year.

1141. Mini Computers. By this year the hand-held computers were in big demand. The eight-digit models
were brought out this year.

1142. Program Controllers. A number of program control-
lers were introduced in 1973. In April, the Modicon
Corporaiton's Model 184 was introduced. This unit
included a read-write memory of up to 4,000 words.
Program controllers were also introduced by Allen
Bradley, Reliance Electric Co. and FX-Systems
Corp.

1143. Computer Printers. Several brands of hard-copy
printers for computers came on the market. The
Xerox 1200 nonimpact printer, using electrophoto-
graphic imagery, was capable of writing 4,000 lines
per minute with 132 characters per line.

1144. Memory. In February, RCA demonstrated the first
holographic optical computer memory to write, store,
read and erase.

INTEGRATED CIRCUITS

1145. IC Fabrication. IC Fabrication. About this time the
Perkin-Elmer Corporation developed the Micralign
Optical Projection printer, a form of noncontact
printer. With this device, yields of integrated circuits were greatly increased, as was the life of the
mask. The system was capable of exposing lines
as narrow as 2 μm with alignment errors of $\pm$ 1 μm.

LASERS

1146. Tuneable Lasers. Bell Laboratories developed a new
class of tuneable laser light sources called fiber
Raman lasers. One of these fiber lasers operated

well over a kilometer in length. These fiber Raman
lasers had a high conversion efficiency in pulsed
or continuous wave modes. Color tuning was possible in the visible spectrum and into the infrared.

1147. Military Lasers. At Kirtland Air Force Base, New
Mexico, a carbon-dioxide laser shot down a drone
aircraft in a Department of Defense test.

PHYSICS

1148. Bedell and Lager. John Bedell and John Wells Lager,
at the Allied Corporation of Morristown, New Jersey, devised the first practical way of producing
metallic glass ribbons.

1149. Giaever, Esaki and Josephson. Ivar Giaever of the
United States, Leo Esaki of Japan and Brian D.
Josephson of Great Britain jointly received the
Nobel Prize in physics for theories advancing the
field of electronics.

POWER

1150. Dworshak Dam. The US Corps of Engineers completed
the Dworshak Dam, the third largest dam in the
United States after the Oroville and Hoover Dams.
The dam, which was 717 feet high and 3,300 feet
long, was located on the north fork of the Clear-
water river in Idaho. In addition to flood control,
the dam provided recreation facilities and irrigation,
as well as power. Three generators provided
400,000 kw. Eventually three more generators
would be installed.

1151. Nuclear Power. A radioisotope power generator of
100 watts output was installed at Ft. Belvoir, Virginia by the US Navy. This was the most powerful
of the thermoelectric generators built to date.
These devices are particularly useful for power in
beacons, weather stations or other devices in remote
locations. They require little or no maintenance and
operate for years.

1152. Nuclear Power Statistics. At the beginning of the
    year there were 27 nuclear power plants in opera-
    tion in the United States, providing about 5 per-
    cent of the power consumed. Fifty-five additional
    plants were under construction at this time.

RADIO ASTRONOMY

1153. Arecibo Observatory. The Arecibo Observatory in
    Puerto Rico was refigured to make it spherical to
    within 0.12". This gave a 100-fold increase in
    sensitivity for radio work and a 1,000-fold increase
    for radar work. It also permitted a radar map of
    Venus to be made as accurate as the then-existing
    lunar radar maps.

RECORDING

1154. Video Recording. A number of new video recording
    and playback systems had been developed by this
    time and several were put on the market for the
    general public. The systems used tapes or discs
    with no standardization between the systems. A
    program recorded on one system could not neces-
    sarily be played back on a competitive system. The
    prices of the units were high and there was little
    demand for them at this time.

SATELLITES

1155. Intelsat 4 F-7. The Intelsat 4 F-7 satellite was
    launched by the United States for the International
    Telecommunication Satellite Organization. A geo-
    stationary orbit point was selected over the equator
    at 30° west longitude. The channel capacity of this
    satellite was 3,000 telephone circuits or 12 television
    channels. The design life was seven years.

SUPERCONDUCTIVITY

1156. Superconductivity. In September, John R. Gavaler,

at the Westinghouse Research Laboratories, discovered that a compound of niobium and germanium became superconducting at 22.3° K. This was the highest temperature at which superconductivity had been observed until later in the year when scientists at Bell Laboratories obtained superconductivity at 23.2° K.

TELEPHONE

1157. <u>California-to-Japan Cable</u>. Approval was given by the FCC for a submarine telephone cable from San Luis Obispo, California to Japan via Hawaii and Guam.

TELEVISION

1158. <u>Aerostat Television Relay</u>. Balloon-borne television transmitters were in use by this time, operating at 10,000 to 15,000 feet altitude. From these elevations areas up to 70,000 square miles had direct line-of-sight reception from the low-power transmitters. One of the leaders in this work was the T-Comm Corporation of Rockville, Maryland. The system was in use in the Bahamas and provided service to all of the islands with television relay from the United States mainland.

1159. <u>Projection Color TV</u>. A seven-foot diagonal television projection system was introduced by the Advent Corporation. The system used three picture tubes with an internal Schmidt optical system. The idea of projection TV grew and continued growing into the 1980's.

<u>1974</u>

AMATEUR

1160. <u>Oscar 7</u>. The seventh amateur satellite to be launched,

Oscar 7, was put in a 115-minute polar orbit on
November 15. Oscar 7 carried a 70-cm and two-
meter up link and a two-meter and ten-meter down
link. A ground command system was included. It
continued operating for over six years. During
this time it transmitted many emergency, medical
and weather bulletins, as well as normal amateur
communications. Weight at the time of launch was
65 pounds.

BROADCASTING

1161. AM-FM Radio Receivers. In June, the US Senate
passed bill #S-585 giving the FCC the authority to
require radio sets costing over $15 to have both
frequency and amplitude modulation band reception
capabilities.

1162. Broadcasting Statistics. By this year, reports indicated that about 85 percent of all cars had radios.
In the largest market areas FM saturation had
reached 90 percent or better. For every person
in the US there were 1.8 radios, with about 40
percent of these having FM capabilities.

1163. Cable TV. By this year, there were about 5,000
cable TV systems in operation in the United States.

COMPONENTS

1164. Gallium Arsenide FETs. The Gallium Arsenide field-
effect transistor was developed this year at Bell
Labs. The high-frequency response of these transistors made them widely used in microwave transmission systems.

COMPUTERS

1165. Microcomputers. The National Semiconductor Corporation of Santa Clara, California brought out its first
16-bit microcomputer on a single board.

1166. Electronic Games. The electronic toy industry had
grown to the point that by the end of the year
more than 20 companies had electronic video games
on the market.

INTEGRATED CIRCUITS

1167. Integrated Optical Circuits. The first integrated optical circuits were developed at the Bell Laboratories
this year. The circuits could contain a laser, modu-
lator, filter, detectors, etc. in a single crystal
structure.

LASERS

1168. Welding. The Ford Motor Co. was the first company
to integrate lasers with their machine tools. The
first use was the welding of automotive underbodies
with a 6 kw CO_2 laser.

1169. Lasers. Using laser techniques, Bell Lab scientists
produced optical pulses less than a trillionth of a
second long.

PHYSICS

1170. Antony Hewish and Martin Ryle. The Nobel Prize in
physics was awarded to Antony Hewish for discov-
ering pulsars and to Martin Ryle for using the
radiotelescope to probe outer space for a high degree of precision.

1171. Metallic Glass. A form of magnetic glass with zero
magnetostriction and with magnetic properties comparable to the best supermalloy was developed at
the Bell Laboratories. The glass was produced in
ribbon form directly from the melt, without the
rolling and heat treating required in the manufacture of supermalloy.

1172. AEC Disbanded. The Atomic Energy Commission
(AEC) was formally disbanded this year. On Octo-

ber 11, President Ford signed into law a bill re-
placing the AEC with a Nuclear Regulatory Commission. The bill also created the Energy Research
and Development Administration (ERDA), which was
to take over the functions of the AEC. It would
also be responsible for the full range of energy
problems.

1173. Atomic Power Statistics. By 1974, 42 atomic plants
in the US were in operation and 183 more were under construction or on order. By this time, atomic
power was substantially cheaper than that from
fossil-fueled plants.

RECORDING

1174. Video Tape Systems. By the year's end there were
at least four companies making video tape recorders.
The best known companies were Sony, Matsushita
Electrical Industrial Co., Toshiba, and Victor. Cas-
settes sold from $13 to $26.

SATELLITES

1175. ATS-6. Regular television broadcasting from an earth
satellite was started this year by NASA's Applications Technology Satellite, ATS-6. A number of
regular TV transmissions were carried this year.

1176. Westar Satellites. In 1973 Western Union received
approval from the FCC to build a commercial satellite system. The satellites developed were known
as Westar and built by the Hughes Aircraft Company. The first Westar, Westar A, was launched
by the US on April 13. Westar B was launched
October 10. The satellites were used for voice,
digital data service and television relay throughout
the United States, Puerto Rico, Hawaii and Alaska.
The satellites carried 12 transponders, each with
a capacity of 600 two-way voice channels or one
television channel.

1177. Weather Satellites. On May 17, the United States

launched SMS-1 to provide continuous pictures of
cloud cover over the US and the Atlantic Ocean.
This was the first synchronous meteorological satellite launched from Cape Kennedy.

SPACE PROBE

1178. Mariner 10. The Mariner 10 spacecraft was launched
from Cape Kennedy, November 3, 1973 and scheduled to pass Venus at about 3,300 miles on February 5. About 45 days later on March 29, Mariner
10 passed within about 460 miles of the surface of
Mercury. After orbiting the sun it again flew by
Mercury on September 21. This probe provided
the first television close-up view of the surface,
which was very rough with craters from many meteorite impacts. The planet Mercury was also discovered as having a weak magnetic field.

1179. Pioneer 2. The spacecraft Pioneer 2 was launched
April 5, 1973, towards Jupiter. On December 2,
the first pictures of the polar regions of Jupiter
were taken as Pioneer 2 flew by at a distance of
about 26,600 miles.

TELEPHONE

1180. Touch-A-Matic Telephone. The first Touch-A-Matic
telephone sets were introduced in 1974. By push-
ing a single button, any prestored number would
be "dialed" into the system. A temporary memory
system provided automatic redialing if desired.
The memory was not lost in case of power failure.

TELEVISION

1181. Television. In 1974, station WLS-TV went on the air
from the Sears Tower in Chicago, the location of
the original Sears broadcasting station WLS. It was
just 50 years from the time station WLS first went
on the air.

1975

AUTOMOBILE ELECTRONICS

1182. Automobile Electronics. Electronic fuel injection was first used in the United States in the Cadillac, built by General Motors.

COMPONENTS

1183. Game Chips. The electronic game industry was becoming a big business. By this time, the General Instrument Company was able to provide integrated circuits for games at a price from $5 to $6.

1184. Tubes. Eimac brought out the 8863 tube.

COMPUTERS

1185. Lois and Richard Heiser. What is reputed to be the world's first store for selling personal computers was opened this year by Lois and Richard Heiser in Santa Monica, California.

INDUSTRY

1186. Piscataway Laboratory. A new laboratory for the development of computer-based information systems was opened by Bell Laboratories in Piscataway, New Jersey.

1187. Gary O'Hara and Tim Shane. TRYOM, Inc., a company formed by Gary O'Hara and Tim Shane, was founded in Beachwood, Ohio to produce electronic games. Their first venture was an electronic backgammon game, which appeared on the market in 1977.

INSTRUMENTATION

1188. Acoustic Microscope. The first acoustic microscope

to hit the market was made by Sonoscan, Inc. of Bensenville, Illinois. Resolution comparable to that of the optical microscope was claimed.

POWER

1189. Solar Power. The Energy Research and Development Administration, in cooperation with the State of New Mexico, started its first solar irrigation project. The Sandia Laboratories and the State University conducted the developments. They reportedly developed a 19-kw solar-powered pump.

1190. Nuclear Power Statistics. In 1975 the US had 55 operating nuclear reactors and another 63 under construction.

SEWING MACHINE

1191. Sewing Machine. The Singer preprogrammed sewing machine was introduced on June 1. It was programmed to perform up to 25 different stitches as well as button holes as set by a selector dial. It was known as the Athena 2000.

SPACE PROBE

1192. Viking Spacecraft. Viking 1 and 2 orbiters made photos of the Martian surface features with pictures having resolution from 0.06 to 0.6 miles. Chemical composition measurements of the atmosphere and soil were made and the results sent to earth.

Viking 1 was launched August 20. The lander touched down on July 20 in the Chryse Planita basin of Mars. This was 17 years to the day after Armstrong and Aldrin landed on the moon.

Viking 2 was launched on September 9, and the lander made a landing on Utopia Planita, about 4,500 miles from Viking 1. Many photographs, soil analyses and other information were relayed by the orbiter to earth.

1193. Venus Photos. The first surface pictures of Venus
were taken by the USSR space probes Venera 9
and 10. Venera 9 was launched June 8, went into
orbit about Venus and launched a capsule that
made a soft landing on the surface on October 22.
It returned pictures of the area and information
for about 53 minutes before failure due to the temperature and pressure. The Venera 10 capsule
landed on October 25 and sent pictures and information back for about 65 minutes before failure.

TELEPHONE

1194. AT&T L5 System. Testing started on the coaxial
cable carrier system known as the L5. When completed, the system utilized a 22-tube coaxial cable
configuration with a capacity of 108.000 two-way
channels.

TELEVISION

1195. Statistics. By the end of 1975 there were 961 television stations on the air, with 351 on the UHF
band and 610 on the VHF band.

TIME

1196. Electronic Watches. By the end of 1975 it was estimated that at least 40 companies were producing
electronic watches with digital displays. Also included in some models were other functions, such
as calendars.

1976

BROADCASTING

1197. TV and Radio Statistics. By the end of 1976 there
were just over 700 commercial television stations in

the US and about 250 public service stations. Approximately 12 million homes were served by cable systems with satellites relaying to about 100 satellite earth stations. About 4,500 AM stations and 3,780 FM stations were in operation at this time.

1198. Federal Communications Commission (FCC). In the early months of the year, it was necessary for the FCC to declare a six-month halt in accepting applications for AM and FM licenses. At that time there was a backlog of over 700 AM and FM station applications.

COMMUNICATION

1199. Citizens Band. The FCC increased the citizens band from 23 to 40 channels in Class D service at about 27 MHz. This ruling became effective January 1, 1977. Over three and a half million applications for CB permits were waiting to be processed by the FCC. In 1976 an estimated five million transceivers were put in service in this country.

COMPONENTS

1200. Fiber Optics Connectors. By the end of the year the Bell Laboratories' engineers had succeeded in the mass production of fiber optic connectors with as many as 144 individual fibers. No individual handling of the fibers was necessary.

1201. Tubes. On April 30, RCA closed down its Harrison, New Jersey receiving tube plant. Westinghouse and Raytheon had previously closed their tube plants. General Electric and Sylvania were now the principal tube manufacturers in the US.

COMPUTERS

1202. Microprocessor Applications. By this time, over two dozen types of microprocessors were on the market.

They were being employed in the control of traffic lights, elevators, electronic cash registers, sewing machines, microwave ovens and automobiles.

ELECTRONIC GAMES

1203. Mostek Corp. By March, the Mostek Corporation brought out an electronic hand-held chess game. The set contained a computer programmed with a chess algorithm into which the player's moves were entered and the defense computed and indicated.

LASERS

1204. Military Laser. In the test of a carbon-dioxide laser mounted on an amphibious vehicle at Redstone Arsenal, both a drone helicopter and aircraft were shot down.

POWER

1205. Nuclear Power. By the end of 1976, and in response to a ruling of the Court of Appeals calling for more consideration of atomic waste management, a hold was called on licensing of nuclear plants. The construction of 11 plants was held up. There were 211 reactors in operation or under construction at this time.

SATELLITES

1206. Marisat. The Marisat communication satellites, built by Hughes Aircraft, were put in operation for improved communication with ships at sea. The system covered the Atlantic, Pacific and Indian Oceans and provided facsimile, teleprinter, and voice communication. Three satellites were launched, designated as Marisat A, B and C, in 1976.

1207. Satellites. Seventeen communication satellites were launched in 1976. Over a hundred countries or

territories on six continents were leasing satellite
services to bring television to over a billion people.

TELEPHONE

1208. Cables. An additional transatlantic cable was installed by the Bell System between the US and
France. The cable permitted 4,200 two-way voice
channels. Repeaters were spaced at 5.1 nautical-
mile intervals.

1209. Digital Switching. The Western Electric digital switching system No. 4ESS was introduced this year.
This system could handle 550,000 toll calls on
107,000 terminations.

TELEVISION

1210. Large-Scale TV. By the end of the year, over two
dozen companies were selling large-screen television
sets. Large-screen sets were selling at prices up
to $5,000.

1211. Texas Instruments. In January, Texas Instruments,
Inc. introduced a five-function electronic watch at
the Consumer Electronics Show. The watch sold for
under $20. This immediately caused other watch-
makers to reduce their prices.

1977

AMATEUR

1212. Harris, Cleveland and Lott. The technique of Narrow
Band Voice Modulation (NEVM) for compressing
speech frequencies to about half the normal bandwidth was developed by Dr. R.W. Harris, J.F.
Cleveland and T. Lott at the University of the
Pacific in Stockton, California. First information
on the system was published in December. The

system permitted more speech channels in a given
communication band.

AUTOMOTIVE ELECTRONICS

1213. Automotive Electronics. General Motors introduced
the use of microprocessor-based engine control systems known as MISAR in the Oldsmobile Toronado.

BATTERIES

1214. Batteries. Gould, Inc. produced a battery for low-
current applications, such as hearing aids and
watches. It had a life of about twice that of the
usually used mercury cells. The silver-mercury
cells were known as the Activair button cells.

COMPONENTS

1215. Pressure-Sensitive Paint. A pressure-sensitive semiconductor paint was announced by Innovations Labs
of Altadena, California. After drying, a single
film provided a no-load resistance of over 100
megohms, which reduced to less than an ohm with
0.45 kg pressure.

1216. Light-Emitting Diodes (LED). By the middle of the
year, the Bell Laboratories were producing light-
emitting diodes (LEDs) with a life of up to a million
hours or more.

1217. Tubes. Industrial Electronic Engineers brought out
the Series DA-2001 and DA2010 replacements for the
RCA digital readout tubes known as Numitrons.

COMPUTERS

1218. Computers. IBM introduced its 3003 processor. Operating speeds were 1.5-1.8 times faster than the
370, the fastest comparable system.

1219. **Electronic Games.** The Gammonmaster, an electronic backgammon game, was introduced at the Consumer Electronics show held by TRYOM Inc.

INSTRUMENTS

1220. **Fluke Mfg. Co.** The Fluke Manufacturing Co. Model 8020 was introduced this year. This was one of the first hand-held multimeters to be widely accepted. It used a liquid crystal display.

INTEGRATED CIRCUITS

1221. **Thomas R. Anthony and Harvey E. Cline.** A new breakthrough in the fabrication of semiconductor devices, known as thermomigration, was invented by Thomas Anthony and Harvey Cline. By this technique the doping of semiconductor materials could be speeded up as much as a thousand times over the older conventional methods.

LASERS

1222. **Lasers.** Semiconductor lasers suitable for optical communication were developed this year in Bell Laboratories.

PATENTS

1223. **R. Gordon Gould.** A patent on the optically pumped laser amplifier was granted to R. Gordon Gould. The patent was based on a notebook memo notarized in November 1957. This had been in litigation for some time.

PHYSICS

1224. **Philip W. Anderson, John H. Van Vieck and Nevill F. Mott.** The Nobel Prize in physics was awarded to Anderson and Van Vieck of the United States and

Mott of the United Kingdom for their work on computer memories and electronic devices.

RADAR

1225. Color Weather Radar. By this time, a new type of
radar was being used in some aircraft. This was
a weather radar with a 186-mile range that indicated heavy rainfall in red, lighter rain in yellow
and a still lighter rain in green. This gave the
pilot a clear indication of the safest path and which
areas were to be avoided while in flight.

1226. Solar Energy. In November, the Reedy Creek Utility
Co. put a solar energy system in operation in an
office building near Orlando, Florida. The system
was to provide space heating and cooling, and hot
water. The collectors replaced the usual roof.
During the first two months after the checking
started, the solar energy system delivered
1,250,000 Btu of space heating. Fuel savings averaged 2.5 million Btu per month.

SPACE

1227. Viking 1. The Viking 1 orbiter offered the first
close view of Phobos, the largest of the moons of
Mars, in its pass by Mars in February. The moon,
only 13 miles in diameter, appeared irregular in
shape.

TELEPHONE

1228. Light Guide Communication. In the Chicago area, the
Bell Telephone Co. installed the first light wave
communication system. Instead of wires, optical
fibers carried the communication--one fiber for each
direction, with each pair carrying 576 voice or data
channels or one 4-MHz video channel.

1229. Electronic Blackboard. The Gemini 100, or electronic
blackboard system, developed by Bell Laboratories,

transmitted drawings, graphs, or handwriting over
the telephone lines and displayed them on a cathode-
ray tube in the remote receiver.

TELEVISION

1230. <u>Pocket TV</u>. A television set with a 2" screen was
put on the market by Sinclair Radionics, Ltd.
The case was 4" x 6" x 1-1/2" and included built-
in antennas for both VHF and UHF channels. The
set operated continuously up to six hours on four
1-1/2-volt rechargeable cells. The set weighed
just over two pounds.

1978

AMATEUR

1231. <u>Oscar 8</u>. On March 5, <u>Oscar 8</u> was launched into a
103-minute polar orbit with a two-meter up link and
ten-meter and 70-cm down links. It was still operating through 1981.

1232. <u>Moon Bounce Test</u>. The first international American
Radio Relay League (ARRL) moon bounce contest
was held April 15-16, and May 20-21. The object
was two-way communication by an earth-moon-earth
path on any authorized amateur frequency over 50
MHz. About one hundred stations participated in
the tests. The best score for single operator stations was made by YV5ZZ, who made 26 contacts
during the test period. YV5ZZ operated on both
the 144 and 432 MHz bands.

1233. <u>Russian Satellite</u>. The amateur satellites RS-1 and
RS-2 were put in orbit this year by the USSR.

AUTOMOTIVE ELECTRONICS

1234. <u>Automotive Electronics</u>. Electronic engine control was

applied to the Lincoln Versailles. Known as the
EEC-1, it controlled ignition timing and the exhaust-
gas recirculation flow-rate. Another system used
for the Pinto and Bobcat to meet California laws
was the type ECU-A.

COMPONENTS

1235. Solar Cells. A tremendous improvement in solar cells
was made this year by the Bell Laboratories. Efficiencies up to 23 percent were achieved.

1236. Electro-optic shutters. Several sizes of electro-optical
shutters were announced by Motorola, Inc., Albu-
querque, New Mexico. Switching time of about one
microsecond could be obtained with these shutters.

1237. Solar Cells. This year at least 10,000 square meters
of single-crystal silicon cells were manufactured.
At 10-percent efficiency, this represented about
one million watts at peak output.

COMPUTERS

1238. Fujitsu and Hitachi. Japan unveiled what they considered to be the largest and fastest computer to
date. Japan appeared to have caught up to the
US in computer techniques. In January, Fujitsu
and Hitachi announced the M-200. The computer
was up to 1.8 times faster than IBM models. Later
in September the M-200H came out. It was about
10 percent faster than the M-200.

1239. Microwave Landing System. This year, the Microwave
Landing System (MLS) based on the Time Reference
Scanning Beam Approach, was agreed upon by the
International Civil Aviation Organization to replace
the Instrument Landing System which had been in
use since 1939. The MLS System was developed
jointly by American and Australian companies.

PHYSICS

1240. Magnetic Bottle. The Doublet III Tokamak, the
world's largest magnetic confinement device for
plasma, was completed and put in operation by the
General Atomic Company in La Jolla, California.
The Tokamak was used for atomic fusion experiments, hopefully leading to fusion-type atomic power plants.

1241. Dr. R. Hasegawa. The magnetic properties and thermal stability of metallic glass were improved by a
group under the direction of Dr. Hasegawa at the
Allied Chemical Company. When used for 60-cycle
transformers, the core loss was only about one
quarter that of silicon steel and well below that of
previous metallic glasses.

1242. Penzias and Wilson. Arno A. Penzias and Robert W.
Wilson of the United States received the Nobel
Prize for their work in cosmic microwave radiation.

POWER

1243. Nuclear Plants. This year was a record year for the
nuclear power industry, with 43 nuclear power
plants scheduled to go on line. Plants in Korea,
Austria, Brazil and Taiwan were scheduled.

1244. Wind Power. An experimental federally sponsored
wind generator was put in service at Clayton, New
Mexico on January 28. The generator was installed
by the Westinghouse Electric Corporation. The cut-
in wind velocity was 9.5 mph. Cut-out occurred
at 40 mph wind velocity. The generator provided
up to 200 kw output in a 22.4 mph wind and could
provide nearly 15 percent of the town requirements
during off-peak periods.

RECORDING

1245. N.V. Phillips. A system of video recording called
Laser Vision was developed in the Netherlands, by

N.V. Phillips of Eindhoven. Recording was on a plastic disc. By means of a turntable it was played back over a conventional television set.

1246. MCA Incorporated. MCA Incorporated of Los Angeles, California released a long-playing video recording system this year. This system used a plastic disc and turntable. Video and audio signals used frequency modulation. The discs played up to two hours.

SATELLITES

1247. Telephone Service. By the end of the year, direct telephone service via satellite had been extended to 124 countries.

TIME

1248. Riehl Time Corporation. A watch manufactured by the Riehl Time Corporation which used silicon cells to absorb energy from ambient light for its operation was being distributed by the Starshine Group of Santa Barbara, California. The watch had been programmed to display the correct month and day for the next 123 years. It would withstand, without damage, being submerged in boiling water for 30 minutes.

1979

AMATEUR

1249. World Administrative Radio Conference (WARC-79). The primary actions taken by WARC-79 of interest to the radio amateurs were those of frequency allocations. Gains resulting for the amateurs included three new amateur bands and exclusive use of parts of the 160- and 80-meter bands previously shared, thus giving 250 KHz more spectrum and 500 KHz more exclusive spectrum use.

1250. <u>Earth-Moon-Earth Communication</u>. The second ARRL
international moon bounce communication contest
was held on April 21-22 and May 19-20, 1979. This
year, 103 stations took part in the tests. The
winner for single-operator stations was K1WHS,
who worked 39 stations on 144 MHz in only one
weekend.

AUTOMOTIVE ELECTRONICS

1251. <u>Automotive Electronics</u>. General Motors introduced
its C-4 closed loop system. The devices computed
the proper fuel and air-flow rates for the motor
and also had self-diagnostic capability, in case of
trouble.

COMMUNICATION

1251a. <u>Fiber Optics</u>. Research in fiber optics had continued
through the decade at the Corning Glass Works,
Corning, New York. By 1979 they were producing
fibers with losses of only 0.7 db/km. Many organizations were using fiber optics for communication
in experimental setups and handling data transmissions at rates of over 100 megabits per second.

COMPONENTS

1252. <u>Tubes</u>. New tubes out this year included: Sylvania
--EY500A/6EC4, PCL805/18GV8, PCL86/14GW8,
ECL86/6GW8, PL504/27GB5, EL504/6GB5, PL519/
40KG6A, EL519/6KG6A (these were primarily re-
placement tubes for television sets); and Siemens
Corp.--Helium Neon gas laser tubes, Models
LGR7621, 7622, 7624, 7627, designed for use in
optoelectronic laser systems.

1253. <u>Nickel-Hydrogen Battery</u>. A new battery was announced in November which was expected to extend
the life of satellites to at least ten years. Under
development at the Ford Aerospace and Communications Corp., the nickel-hydrogen battery would replace the nickel-cadmium types.

1254. Zinc-Nickel Oxide Battery. The Delco-Remy division of General Motors developed a zinc-nickel oxide battery about half the weight and size of the lead-acid batteries used in motor vehicles in 1979.

1255. Sodium-Sulphur Battery. A very unusual form of battery was announced. This battery utilized liquid electrodes, sodium and sulphur, and a solid electrolyte and separator beta-alumina, a ceramic material conductive to sodium ions. The reaction created Na_2S_3. Because active ingredients must operate in the molten state, the battery could only operate at a temperature between 300° and 350° C. The current collector was graphite.

1256. Ford. By 1978 it was determined that in order to bring cars to more efficient operation with cleaner exhausts, a small computer system would be necessary. This year, the Ford Motor Co. brought out the EEC-11 control for its cars. It controlled the timing, exhaust recirculation and air/fuel ratio as required for optimum operation.

COMPUTERS

1257. Light-Powered Calculator. The first light-powered calculator was announced this year. The calculator had no batteries and operated on ambient, fluorescent or incandescent light.

CONFERENCES

1258- World Administrative Radio Conference (WARC-79).
9. WARC-79 was scheduled to begin on September 24 and run until November 20 in Geneva, Switzerland. This was the first international meeting since 1959. Delegates from 142 countries attended. About 14,000 proposals had to be considered and the conference did not close until December 4, 1979.

INSTRUMENTATION

1260. Tektronix Oscilloscope. Tektronix produced the first

nonsampling oscilloscope model to provide a 1-GHz
bandwidth. It was designated as Model 7104.

PHYSICS

1261. <u>Magnetic Bearings</u>. Practical magnetically supported
bearings were announced this year. Electronically
controlled shafts weighing 1,100 Kg were rotated
at up to 10,000 RPM or faster with no mechanical
contact. A number of such bearings are said to
be in use at the present time (1984) for grinding,
polishing, and other applications.

1262. <u>Magnetic Levitation Transportaiton</u>. The Thyssen
Henschel Company in Kassel, West Germany demonstrated a "maglev" system with monitoring and
traffic control at the International Transport Exhi-
bition in Hamburg this year.

1263. <u>Alan S. Willsky</u>. Alan S. Willsky received the Nobel
Prize for his paper "Relationship Between Digital
Signal Processing and Control and Estimation The-
ory" published in the <u>IEEE Proceedings</u>.

POWER

1264. <u>Solar Power</u>. In December, the Israeli government
put in operation a solar power plant at Ein Bokek.
The experimental system was set up to test the
theory of solar pond power conversion.

1265. <u>Nuclear Power Statistics</u>. By the end of this year
there were 132 nuclear power plants in operation
around the world with 223 reactors. There were
97 new plants under construction.

RECORDING

1266. <u>Laser-Recording</u>. The world's first laser optical recording system was demonstrated by Philips Data
Systems in the Netherlands. 10^{10} bits, or the
equivalent of one half of a million typewritten pages

could be put on a 12-inch disc. Any address could
be reached in 250 ms.

SPACE

1267. Voyager 1. In March, the Voyager 1 space probe
vehicle flew by Jupiter and returned photographs
of the four largest moons of Jupiter: Callisto,
Ganymede, Europa and Io. These photographs
also showed rings of space debris which had not
been detected previously.

1268. Voyager 2. In July, Voyager 2 flew through the region, giving additional views of Jupiter's moons,
rings and the volcanic eruptions on Io.

1269. Pioneer 2. Pioneer 2 (launched in April 1973) passed
within 26,600 miles of Jupiter in 1974 and in December of this year gave the first pictures of the polar
regions of Jupiter. By September, Pioneer 2
reached Saturn, giving new views of the rings and
Titan, the only known moon to have a dense atmosphere.

TRANSPORTATION

1270. Electric Car. The Department of Energy introduced
an electric test car with a cruising speed of 55
mph while carrying four passengers. It was developed by the R & D department of General Electric Co. in Schenectady, New York. Mass production is expected to begin about 1985.

1271. Transportation. Magnetic levitation was demonstrated
on a vehicle and operated at the international
Transport Show in Hamburg, Germany. Ninety-
three kilometers per hour over a 900-meter route
was obtained.

BIBLIOGRAPHY

Many of the references listed in the first volume of Electrical and Electronic Technologies also provide valuable information on the period covered by this volume. Additional references which have been of value in preparing this and the second volume are listed below.

BOOKS

Austin, John Benjamin. Electric Arc Welding. Chicago: American Technical Society, 1952.

Baxter, James Phinney, III. Scientists Against Time. Boston: Little, Brown and Company, 1948.

Beck, A.H.W. Words and Waves. New York: World University Library/McGraw-Hill Book Company, 1967.

Bell Laboratories Innovations in Telecommunications 1925-1927, by Staff of Bell Labs. Murray Hill, N.J., and New York, 1979.

Bergmann, Ludwig. Ultrasonics. New York: John Wiley & Sons, 1938.

Berkeley, Edmund C. Giant Brains or Machines That Think. New York: John Wiley & Sons, 1949.

The Book of Popular Science. New York: Grolier, Inc., 1963.

Brooks, J. Telephone, the First Hundred Years. New York: Harper & Row, 1976.

Brown, Ronald. Lasers--Tools of Modern Technology. Garden City, N.Y.: Doubleday & Co., 1968.

_____. Telecommunications. Garden City, N.Y.: Doubleday & Co., 1970.

Brown, Sanborn C. Count Rumford Physicist Extraordinary. Garden City, N.Y.: Doubleday & Co., 1962.

Bucher, Elmer C. Practical Wireless Telegraphy. New York: Wireless Press, 1917.

Casson, Herbert N. The History of the Telephone. Chicago: A.C. McClure, & Co., 1910.

Chalmers, J.A. Atmospheric Electricity. Long Island City, N.Y.: Pergamon Press, 1957.

Chaloner, W.H. People & Industries. London: Frank Cass & Co., Ltd., 1963.

Cook, J. Gordon. Electrons Go to Work. New York: The Dial Press, 1957.

Cowan, H.J. Time and Its Measurement. Cleveland: World Publishing Co., 1958.

Crawley, Chetwode. From Telegraphy to Television. New York: Frederick Warne & Co., 1931.

Crowther, J.G., and R. Whiddington. Science at War. New York: Philosophical Library, 1948.

Culver, Charles A. Electricity and Magnetism. New York: The Macmillan Co., 1930.

Dawson, Frank. Nuclear Power. Seattle: University of Washington Press, 1976.

DeForest, Lee. Father of Radio: An Autobiography of Lee DeForest. New York: Wilcox and Follett Co., 1950.

Dreher, Carl. Sarnoff, an American Success. Chicago: Quadrangle Books, 1977.

Encyclopaedia Britannica. Eleventh Edition. Chicago: Encyclopaedia Britannica Press, 1911.

Encyclopedia Americana Year Books, 1958-1979. New York: Americana Corp., 1958- .

Fessenden, Helen M. Builder of Tomorrows. New York: Coward-McCann, Inc., 1940.

Fleming, J.A. The Thermionic Valve and Its Developments in Radio Telegraphy and Telephony. Sydney, N.S.W.: Wireless Press, Ltd., 1919.

Garrison, G.R. Radio and Television. New York: Appleton-Century-Crofts, Inc., 1950.

Ginzburg, Benjamin. The Adventure of Science. New York: Tudor Publishing Co., 1932.

Goldsmith, N. Radio Telephony. New York: The Wireless Press, Inc., 1918.

Greenwood, Ernest. Aladdin USA. New York: Harper and Brothers Publishers, 1928. [About Thomas A. Edison].

Haga, Enoch. Understanding Automation. Elmhurst, Ill.: The Business Press, 1965.

Hammond, John Winthrop. Charles Proteus Steinmetz. New York: The Century Co., 1924.

Hay, J.S. Evolution of Radio Astronomy. New York: Science History Publications, 1973.

Hazeu, H.A. Fifty Years of Electronic Components, 1921-1971. Eindhoven, Netherlands: N.V. Phillips, Gloeilampenfabrieken, 1971.

Head, S.W. Broadcasting in America. Boston: Houghton Mifflin Co., 1976.

Heiserman, D. Radio Astronomy. Blue Ridge Summit, Pa.: Tab Books, 1975.

Houston, Edwin J. Electricity in Every-Day Life, Vol. III. New York: P.F. Collier & Son, 1905.

Howeth, Capt. L.S. History of Communications--Electronics

in the United States Navy. Washington, D.C.: U.S. Govt. Printing Office, 1963.

Ingels, Margaret. Father of Air Conditioning. Garden City, N.Y.: Country Life Press, 1952.

Jespersen, James, and Jane Fitz-Randolph. Time and Clocks for the Space Age. New York: Atheneum, 1979.

Josephson, Matthew. Edison. New York: McGraw-Hill Book Co., Inc., 1959.

Lessing, Lawrence. Man of High Fidelity: Edwin Howard Armstrong. New York: Grosset & Dunlap, 1969.

Lichty, Lawrence W., and Malachi C. Topping. A Source Book on the History of Radio and Television. New York: Hastings House Publishers, 1975.

Lincoln, E.S. Industrial Electric Lamps and Lighting. New York: Essential Books, 1945.

Livingston, M.S. Particle Accelerators: A Brief History. Cambridge, Mass.: Harvard University Press, 1969.

Lyons, Eugene. David Sarnoff. New York: Harper & Row, 1976.

Mackenzie, Catherine. Alexander Graham Bell: The Man Who Contracted Space. Boston: Houghton Mifflin Co., 1928.

Mann, Martin. Revolution in Electricity. New York: Viking Press, Inc., 1962.

Manson, Arthur J. Railroad Electrification and the Electric Locomotive. New York: Simmons Boardman Publishing Co., 1925.

Meissner, Benjamin Franklin. On the Early History of Radio Guidance. San Francisco: San Francisco Press, 1964.

Mills, John. Radio Communication Theory and Methods. New York: McGraw-Hill Book Company, 1917.

Moseley, Mabuth. Irascible Genius: The Life of Charles Babbage. Chicago: Henry Regnery Co., 1964.

Nausmann, Eric, et al. Radiophone Receiving. New York: D. Van Nostrand Co., Inc., 1922.

Novick, S. The Electric War: The Fight over Nuclear Power. San Francisco: Sierra Club Books, 1976.

Piddington, J.H. Radio Astronomy. New York: Harper & Row, 1962.

Poole, L. and G. Electronics in Medicine. New York: McGraw-Hill Book Co., 1964.

Pupin, Michael. From Immigrant to Inventor. New York: Charles Scribner's Sons, 1930.

Rolt-Wheeler, Francis. Thomas Alva Edison. New York: The Macmillan Company, 1922.

Sheldon, Samuel, and Erich Hausmann. Electric Traction and Transmission Engineering. 2nd ed. New York: D. Van Nostrand Co., Inc., 1920.

Shippen, Katherine B. Mr. Bell Invents the Telephone. New York: Random House, 1955.

Slattery, Harry. Rural America Lights Up. Washington, D.C.: National Home Library Foundation, 1940.

Southworth, G.C. Forty Years of Radio Research. New York: Gordon and Breach Science Publishers, Inc., 1962.

Sullivan, George. Rise of the Robots. New York: Dodd, Mead & Company, 1971.

Tate, A.O. Edison's Open Door. New York: E.P. Dutton & Co., 1938.

Upton, M. Electronics for Everyone. New York: Devin-Adair Co., 1957.

Van der Bijl, N.J. The Thermionic Vacuum Tube. New York: McGraw-Hill Book Co., 1920.

Viemeister, Peter E. The Lightning Book. Garden City,
N.Y.: Doubleday & Company, 1961.

Vinal, G.W. Storage Batteries. New York: John Wiley &
Sons, 1924.

War Department Document No. 1069. The Principles Under-
lying Radio Communication. Washington, D.C.: Government Printing Office, 1922.

Weedy, B.M. Electric Power Systems. New York: John
Wiley & Sons, 1972.

Williams, Henry Smith. The Story of Modern Science, Vol.
IX. New York: Funk & Wagnalls Co., 1923.

PAMPHLETS

Anderson, Leland. Priority in the Invention of Radio: Tesla vs. Marconi. Holcomb, N.Y.: Antique Wireless Association Monograph #4, undated.

Sivowitch, Elliot N. A Technological Survey of Broadcasting's "Pre-History," 1876-1920. Holcomb, N.Y.: Antique Wireless Association, Monograph #2, undated.

Some RCA "Firsts" in the Radio World. New York: NBC
Department of Information, undated.

Staff-WHA. The First 50 Years of University of Wisconsin
Broadcasting, WHA, 1919-1969. Madison: University of
Wisconsin, undated.

Warner, J.C., et al. Five Historical Views. History of RCA
to 1971. New York: RCA, undated.

The Zenith Story. Chicago: Zenith Radio Corporation,
1955.

ARTICLES

Angus, Robert. "75 Years of Magnetic Recording," High
Fidelity, March 1973.

Armstrong, Edwin H. "Frequency Modulation," Electrical Engineering, December 1940.

Ayer, D.R., and C.A. Wheeler. "The Evolution of Strip Transmission Line," Microwave Journal, May 1969.

Black, Harold S. "Inventing the Negative Feedback Amplifier," IEEE Spectrum, December 1977.

Danilou, Dr. Victor J. "America's Greatest Discoveries, Inventions, and Innovations," Industrial Research, November 15, 1976.

"Light from Under the Bushel," Canadian Consulting Engineer, February 1963. [On the incandescent lamp developments of Henry Woodward].

Oudin, J.M., and R.A. Teller. "Submarine D.C. Cables." IEEE Spectrum, July 1966.

Smith, Desmond. "Radio Comes on Strong," Electronic Age, Autumn 1963.

APPENDIX I: INDEX TO VACUUM TUBES

TUBE TYPE	DATE	ENTRY NUMBER
Cermolox	1960	689
Compactron	1961	725
Haydu	1951	401
Kalotron	1953	461
Klystron hi-power	1955	520
Loktal	1945	160
Metal picture tube	1948	304
Microtubes	1941	53, 520
Nixie	1951	401, 574
Noise tube	1949	332
Nuvistor	1959	652, 728, 810
Phasitron	1946	203
Plasmatron	1951	400
Proximity Fuse	1941	52
Stacked	1954	491
T-R	1941	54
TWT hi-power	1945	161
Vibrotron	1946	211
Vidicon	1949	330, 365
OA3/VR75	1944	130
OA5	1947	261
OC3/VR105	1944	130
OD3/VR150	1944	130
C1b/A	1949	331
TGC-1	1948	302
TT-1	1948	302
1AC5	1949	331
1AD4	1949	331
1AD5	1949	331
1AE5	1949	331
1AF4	1950	362
1AF5	1950	362
1B3GT	1949	331, 491
1B35	1946	212
1B37	1946	212
1B48	1945	159
1B67	1949	331
1B85	1949	331
1B87	1949	331
1B90	1949	331

TUBE TYPE	DATE	ENTRY NUMBER
1C	1949	331
1C3	1949	331
1DB	1940	12
1E8	1949	331
1L6	1949	331
1LB4	1940	12
1LC5	1940	12
1LC6	1940	12
1LD5	1940	12
1LG5	1946	212
1P28	1944	130
1P37	1947	261
1P42	1947	261
1R4/1294	1943	112
1R5	1940	12
1S5	1940	12
1T4	1940	12
1T5	1940	12
1T6	1949	331
1U5	1947	261
1U6	1950	362
1V2	1950	362
1W4	1949	331
1X2	1947	261, 362
1X2B	1954	491
1Y2	1947	261
1Z2	1945	159
TGC-2	1948	302
2AF4A	1954	492
2B23	1945	159
EE2BT	1945	159
2C21	1942	86
2C33	1943	112
2C34/RK34	1943	112
2C39	1946	212
2C39A	1950	362
2C40	1945	159
2C43	1946	212
2C44	1944	130
2CW4	1963	810
2D21	1944	130
2DS4	1963	810
2E25	1946	212
2E26	1947	261
2E30	1946	212
2E31	1946	212
2E32	1945	159
2E35	1946	212
2E36	1945	159

TUBE TYPE	DATE	ENTRY NUMBER
2E41	1946	212
2E42	1945	159
2G21	1946	212
2G22	1945	159
2K26	1948	302
2X26	1947	261
3-25A3	1944	130
3-25D	1944	130
3-400Z	1961	729
3-1000Z	1961	729
3AL5	1954	492
3AU6	1954	492
3AV6	1954	492
3B4	1947	261
3B7/1291	1943	112
3B24W	1951	399
3B25	1944	130
3BZ8	1945	159
3B29	1951	399
3BP1	1943	112
3BY6	1954	492
3C	1949	331
3C22	1944	130
3C24/24G	1944	130
3C28	1946	212
3CF6	1954	492
3CX10,000A7	1961	729
3D6/1299	1943	112
3D23	1946	212
3D24	1947	261
3D26	1946	212
AV3E	1947	261
3E5	1950	362
3EP1/1806-P	1943	112
AV3K	1947	261
3KP11	1949	331
3LF4	1942	86
3Q4	1941	55
3QP1	1948	302
3RP1	1949	331
3S4	1941	55
3SC5	1948	302
3X100A11/2339	1945	159
3X2500A3	1945	159
3X3000F1	1955	520
3Y125,000A3	1947	261
4-65A	1947	261, 331
4-100	1951	399
4-125A	1945	159

TUBE TYPE	DATE	ENTRY NUMBER
4-150A	1948	302, 331
4-250A	1945	159
4-250A/5D22	1949	331
4-400A	1947	261
4-400C	1972	1106
4-500A	1947	261
4-750A	1947	261
4BQ7	1954	492
4BZ7	1954	492
4C32	1946	212
4C35	1946	212
4C36	1946	212
4CV1500B	1966	923
4CV100,000C	1967	952
4CV250,000C	1967	952
4CW800B	1970	1050
4CW800F	1970	1050
4CW50,000	1969	1016
4CW100,000E	1969	1016
4CX250K	1956	549
4CX300A	1956	549
4CX600B	1970	1050
4CX600F	1970	1050
4CX600J/8809	1970	1050
4CX1000A	1957	574
4CX1500A	1966	923
4CX1500B	1966	923
4CX3000A	1962	762
4CX5000A	1955	520
4D21	1947	261
4D21/4-125A	1950	362
4D22	1946	212
4D32	1946	212
4D250A/5D22	1950	362
4E27A/5-125B	1949	331
4KM70SJ	1962	762
4SN1A1	1947	161
4W300B	1953	461
4W1250A	1949	331
4W20,000A	1951	399
4X150D	1952	430
4X150G	1948	302
4X250B	1955	520
4X250F	1955	520
4X250M	1955	520
5-500A	1966	923
5AQ5	1954	492
5AS8	1954	492
5C24	1946	212
5CX1500A	1966	923

TUBE TYPE	DATE	ENTRY NUMBER
5D24	1946	212
5DC-5	1953	461
5DC-5M	1953	461
5RP	1945	159
5SP	1945	159
5U4GB	1945	159, 491
5WP15	1948	302
5X8	1954	492
5XP	1949	331
5Y3/5Y3G	1941	55
6AB4	1949	331, 362
6AB5	1950	362
6AB5/6N5	1940	12
6AD4	1950	362
6AG5	1949	331
6AH6	1949	331
6AH7GT	1941	55
6AJ4	1952	430
6AK5	1942	86, 130, 159, 549
6AK6	1944	130
6AL5	1944	130, 331
6AL6G	1940	12
6AL7GT	1950	362
6AM4	1952	430
6AN5	1949	331
6AQ6	1944	130
6AR5	1948	302
6AS5	1948	302, 362
6AT6	1945	159
6AU5	1950	362
6AU7	1954	492
6AV6	1948	302
6AX4GT	1950	362
6AX5	1950	362
6B6GA	1955	520
6BA6	1945	159
6BA7	1948	302
6BC5	1950	362
6BE6	1945	159
6BF5	1950	362
6GB6G	1949	331
6BH6	1948	302
6BJ6	1947	261
6BK5	1952	430
6BK6	1949	331
6BN6	1955	520
6BQ6	1947	261, 362
6BQ6GT	1949	331, 362
6BQ6GTA	1954	491

TUBE TYPE	DATE	ENTRY NUMBER
6BT6	1949	331
6BU6	1949	331
6BX7GT	1952	430
6C	1949	331
6CS6	1955	520
6CW4	1960	728
6CZ4	1945	159
6D22	1946	212
6DS4	1963	810
6E5	1950	362
6GJ5	1962	762
6GK5	1963	810
6J4	1944	130
6J6	1949	331
6K6GT	1949	331
6L6GB	1955	520
6L6WGB	1950	362
6N4	1945	159
6S4A	1954	492
6SA7	1948	302
6SF7	1941	55
6SG7	1941	55
6SJ7GT	1948	302
6SK7GT	1948	302
6SL7	1941	55
6SL7W	1950	362
6SN7GT	1941	55, 549
6SN7GTA	1954	491
6SN7GTB	1954	492
6SN7W	1950	362
6SQ7GT	1948	302
6SS7	1941	55
6T8	1947	261
6U4GT	1950	362
6U5	1950	362
6U8	1952	430
6W4	1948	302
6W4GT	1948	302, 520
6W6GT	1950	362
6X5WGT	1950	362
7AG7	1946	212
7B4	1940	12
7B5	1939	331
7C4/1203A	1943	112
7C5	1938	331
7C29	1946	212
7CP1/1811-P1	1943	112
7E5/1201	1943	112

TUBE TYPE	DATE	ENTRY NUMBER
7EP4	1946	212
7F7	1949	331
7F8	1947	261
7G7/1232	1940	12
7H7	1940	12, 331
7J7	1940	12
7JP4	1949	331
7L7	1940	12
7N7	1949	331
7V7	1941	55
7W7	1942	86
7X6	1949	331
7Z4	1949	331
8AP4	1949	331
8BP4	1949	331
9C21	1944	130
9C22	1944	130
9C24	1949	331
9C28	1947	261
9C29	1947	261
9C30	1947	261
9C31	1947	261
10FP4	1947	261
10KP7	1949	331
12A6GT	1943	112
12AH7GT	1941	55
12AL5	1947	261
12AT6	1945	159
12AT7	1947	261, 362
12AU7	1947	261
12AV6	1948	302
12AX4	1954	492
12AX7	1947	261
12AY7	1949	331, 362
12BA6	1945	159
12BA7	1948	302
12BE6	1945	159
12BH7A	1954	492
12BK6	1949	331
12BQ6GTB	1954	492
12BT6	1949	331
12BU6	1949	331
12BY7A	1954	492
12GH7	1950	362
12H6	1941	55
12K8	1940	12
12KP4	1949	331
12S8GT	1949	331
12SA7GT	1948	302

TUBE TYPE	DATE	ENTRY NUMBER
12SF7	1941	55
12SG7	1941	55
12SJ7GT	1941	55, 302
12SK7GT	1948	302
12SL7GT	1941	55
12SN7GT	1941	55
12SQ7GT	1948	302
12SR7	1940	12
12SR7GT	1943	112
12W6GT	1954	492
13CW4	1963	810
14F8	1947	261
14S7	1942	86
15E	1944	130
16AP4	1949	331
16GP4	1949	331
17CP4	1951	399
HV18	1946	212
19J6	1948	302
19T8	1947	261
TUF20	1945	159
KU23	1946	212
25T	1944	130
25BQ6	1947	261, 362
25BQ6GT	1949	331
25CD6GA	1954	492
26BK6	1949	331
HK27	1947	261
28D7W	1950	362
CE29	1944	130
HY30Z	1940	12
35TG	1946	212
TB35	1946	212
35W4	1945	159
35Z3	1942	86
35Z6G	1940	12
TR40M	1945	159
45Z3	1941	55
50B5	1945	159
50C5	1948	302
50C6G	1940	12
53A	1944	130
55J6	1950	362
HK57	1947	261
HD59	1946	212
RK61	1947	261
HY65	1941	55
RK65	1940	12
HY67	1941	55

TUBE TYPE	DATE	ENTRY NUMBER
V70D	1946	212
70L7GT	1940	12
75TH	1944	130
HY75	1940	12
HY75A	1947	261
TW75	1943	112
VR75-30	1940	12
ECL86/6GW8	1979	1252
PCL86/14GW8	1979	1252
117M7GT	1940	12
117P7GT	1941	55
117Z3	1946	212
127A	1944	130
VXR130	1949	331
152TH	1944	130
152TL	1940	12
PL172	1956	549
207M	1946	212
Z225	1941	55
227A	1944	130
233	1945	159
HK257	1940	12
257B	1950	362
257C	1947	261
GH302	1947	261
GG304	1947	261
304T	1944	130
304TL	1940	12
CE306	1944	130
327A	1944	130
AT340	1946	212
357C	1947	261
371B	1951	399
446A	1944	130
446B	1944	130
450TL/HK854L	1950	362
492R	1948	302
492R/5758	1950	362
EY500A/6EC4	1979	1252
502A	1946	212
Y503	1970	1050
EL504/6GB5	1979	1252
PL504/27CB5	1979	1252
CK505AX	1946	212
CK510X	1945	159
Y518	1970	1050
EL519/6KG6A	1979	1252
PL519/40KG6A	1979	1252
527	1944	130

TUBE TYPE	DATE	ENTRY NUMBER
WL530	1945	159
SD568	1965	891
571AX	1948	302
UE572A	1963	810
572B	1965	891
572B/T-160L	1967	952
576	1951	399
577	1951	399
578	1951	399
592	1946	212
592/3-200A3	1949	331
599	1944	130
X602	1956	549
CK605CX	1947	261
CK608CX	1947	261
NL614	1949	331
HY615	1940	12
616J	1949	331
617	1948	302
635	1950	362
643	1951	399
653	1949	331
672	1948	302
678	1945	159
705A	1953	461
706Y	1940	12
707A	1940	12
710	1947	261
714	1947	261
715C	1949	331
V801/UX41A	1949	331
803	1944	130
PCL805/18GV8	1979	1252
811	1940	12
811A	1963	810
812	1940	12
812H	1946	212
815	1940	12
816	1941	55
822S	1945	159
825	1940	12
826	1941	55
828	1940	12
829	1940	12
829-B	1944	130
833A	1946	212
838	1946	212
849	1946	212
866A/866	1941	55

TUBE TYPE	DATE	ENTRY NUMBER
880	1949	331
889A	1946	212
891	1946	212
892M	1946	212
893A	1945	159
928	1940	12
949A	1946	212
949H	1946	212
1005	1942	86
1006	1943	112
1018	1949	331
1019	1949	331
1080	1945	159
1090	1945	159
R1130	1946	212
1201	1942	86
1203	1942	86
1204	1942	86
1269	1942	86
1291	1942	86
1293	1942	86
1457	1942	86
1614	1950	362
1625	1941	55
1626	1941	55
1628	1940	12
1631	1941	55
1632	1941	55
1633	1941	55
1634	1941	55
1847	1940	12
1945	1947	261
1946	1947	261
1947	1947	261
1950	1947	261
DA2001	1977	1217
DA2010	1977	1217
Z2061	1951	399
X2159	1971	1080
X2170	1973	1139
X2176	1973	1139
L3211	1959	651
R4300	1947	261
R4340	1947	261
4604	1965	891
F5303	1945	159
F5512	1950	362
5513	1949	331
5516	1947	261

TUBE TYPE	DATE	ENTRY NUMBER
5527	1947	261
5544	1947	261
5545	1947	261
5563	1947	261
5594	1947	261
5610	1949	331
5618	1948	302
5630	1947	261
5645	1950	362
5646	1950	362
5648	1947	261
5651	1948	302
5652	1948	302
5653	1947	261
5654	1949	331
5658	1948	302
5663	1947	261
5665	1947	261
5671	1949	331
5674	1974	261
5675	1950	362
5690	1953	461
5691	1947	261, 302
5692	1947	261, 302
5693	1947	261, 302
5696	1948	302
5702/CK605CX	1948	302
5703/CK608CX	1948	302
5704/CK606BX	1948	302
5713	1948	302
5719	1953	461
5722	1949	331
5734	1949	331
CK5744/CK619CX	1949	331
5762	1949	331
5763	1949	331
5770	1949	331
5771	1949	331
5786	1949	331
5794	1949	331
5803/VX34	1949	331
5819	1949	331
5823	1949	331
5825	1949	331
5826	1950	362
5829	1950	362
5831	1950	362
5840	1953	461
5841	1950	362

TUBE TYPE	DATE	ENTRY NUMBER
5844	1950	362
5851	1950	362
5855	1950	362
5893	1957	574
5894/8737	1967	952
5899	1953	461
F5918	1950	362
5946	1950	362
6019	1950	362
6021	1953	461
6039	1951	399
6134	1953	461
6146	1952	430, 729
6146A	1961	729, 848
6146A/8298A	1964	848
6202	1953	461
6203	1953	461
6252	1953	461
6549	1955	520
6569	1955	520
6883B	1964	848
7034/4X150A	1957	574
7270	1960	690
7360	1959	652
7378	1965	891
7586	1961	728
7587	1961	728
LGR7621	1979	1252
7622	1979	1252
7624	1979	1252
7627	1979	1252
7815AL	1970	1050
7835	1961	726
7895	1963	810
7963	1963	810
8001	1941	55
8002R	1940	12
8005	1941	55
8010R	1942	86
8032	1962	762
8032A	1964	848
8056	1963	810
8058	1963	810
8070	1962	762
8072	1962	762
8072W	1971	1080
8117	1962	762
8118	1964	848
8121	1962	762

TUBE TYPE	DATE	ENTRY NUMBER
8121W	1971	1080
8122	1962	762
8122W	1971	1080
8163	1965	891
8179	1963	810
8186	1963	810
8298	1964	848
8300	1962	762
8348	1964	848
8408	1964	848
8457	1965	891
8458	1965	891, 923
8462	1965	891
8505	1964	848, 891
8552	1964	848
8579	1965	891
8583/267	1966	923
8637	1965	891
8828W	1971	1080
8847	1970	1050
8863	1975	1184
8873	1971	1080
8874	1971	1080
8875	1971	1080
8877	1971	1080
8938	1972	1106
9001	1941	55
9002	1941	55
9003	1941	55
AX9900/5866	1950	362
AX9901/5867	1950	362
AX9902/5868	1950	362
AX9903/5894	1950	362, 399
AX9906/6078	1950	362
AGR9950/5869	1950	362
AX9951/5870	1950	362

APPENDIX II: INDEX TO RADIO STATIONS

STATION	ENTRY NUMBER		STATION	ENTRY NUMBER
WCBS-TV	288		WTMJ-FM	151
WCRB	1011, 1032		WTTG	247
WGBH	1011, 1032		WWV	43, 78, 106,
WKCR	1011, 1032			127, 293,
WLS-TV	1181			318, 354,
WMFM	108, 151			391, 946
WNYC	1011, 1032		WWVH	318, 355, 391
WOL	128		WWVL	714